The Fundamental
Constants
A Mystery of Physics

Harald Fritzsch • Translated by Gregory Stodolsky

THE FUNDAMENTAL
CONSTANTS
A MYSTERY OF PHYSICS

World Scientific

NEW JERSEY · LONDON · SINGAPORE · BEIJING · SHANGHAI · HONG KONG · TAIPEI · CHENNAI

Published by

World Scientific Publishing Co. Pte. Ltd.

5 Toh Tuck Link, Singapore 596224

USA office: 27 Warren Street, Suite 401-402, Hackensack, NJ 07601

UK office: 57 Shelton Street, Covent Garden, London WC2H 9HE

Library of Congress Cataloging-in-Publication Data
Fritzsch, Harald, 1943–
    [Absolut Unveränderliche. English]
    The fundamental constants : a mystery of physics / by Harald Fritzsch ;
translated by Gregory Stodolsky.
        p. cm.
    Includes index.
    ISBN-13: 978-981-281-819-5 (hardcover : alk. paper)
    ISBN-10: 981-281-819-7 (hardcover : alk. paper)
    ISBN-13: 978-981-283-432-4 (pbk. : alk. paper)
    ISBN-10: 981-283-432-X (pbk. : alk. paper)
    1. Physical constants. I. Title.
    QC39.F7613 2009
    530.8'1--dc22

                                                    2009001806

British Library Cataloguing-in-Publication Data
A catalogue record for this book is available from the British Library.

This English edition is a translated version of the German edition *Das Absolut Unveränderliche*,
© 2005 by Piper Verlag GmbH, München.

Cover image: Inside a Radio Telescope, © Roger Ressmeyer/CORBIS.

Printed in Singapore.

PREFACE

It was already an important question to the philosophers of antiquity: of what does matter fundamentally consist? When repeatedly dividing some piece of matter such as wood or metal or a diamond, does one reach a limit? If so, how does this limit manifest itself? Are there indivisible objects; is there a smallest possible object? Or does the limit exist only in the sense that any further division would seem to make no sense or be experimentally impossible?

Every careful observer of natural phenomena is bound to be impressed by the colorful and stunning diversity of the material world. One soon notes, however, that there does not exist total chaos in these phenomena. Things repeat themselves. A diamond here and another in some other place are as alike as peas in a pod. The leaves of an oak in Boston are indistinguishable from those of an oak in Denver.

So, within the uncountable diversity of phenomena, there are also constants, things that repeat themselves. It was this duality of multiplicity and constancy that inspired the Greek philosophers, most notably Leucippus of Milet and his student Democritus of Abdera in the 5th century BC, to formulate the hypothesis that the universe is made up of many small, indivisible building blocks called atoms (derived from the Greek word *atomos*, which means essentially the same thing as indivisible). A small number of different atoms and their unending new combinations should suffice, so they claimed, to make up the diversity of things. "Nothing exists," spoke Democritus, "except atoms and empty space."

A slightly different concept was brought into play by Anaxagoras around 500 BC. He spoke of an infinity of basic elements which, by mixing, would produce the diversity of objects in the world. These basic elements were furthermore said to be indestructible, and the changes observable in physical objects were considered the result of motion causing new combinations of the elements. Empedocles, who was ten years younger than Anaxagoras, postulated that the basic materials of the world were the four elements earth, water, air, and fire.

It is interesting to note that it is in the conceptualization of the atom that empty space plays a role for the first time. Up to this point, space had been seen as filled with matter, and the idea of empty space was therefore unthinkable. In the context of the theory of the atom, empty space took on an important function. It became the bearer of geometry, the structure in which the atoms moved. So now matter and geometry are two different things.

Atoms move in space and they have geometrical qualities. Democritus said: "Just as tragedy and comedy can be written down with the same letters, different phenomena can be realized in our world by the same atoms, provided they assume different positions and different movements. Some given matter may have the appearance of a given color; may appear to us as tasting sweet or bitter — but in reality there are only atoms and empty space."

Later the Greek philosophy adopted the elements of the theory of the atom and developed the idea further. Plato, in his dialogue *Timaeus*, discusses possible connections between atoms and the Pythagorean theory of the harmony of numbers. For instance, he identified the atoms of the elements earth, water, air, and fire with the regular solids, the cube, octahedron, icosahedron, and tetrahedron. In speaking of the motion of the atoms, special reference is made to natural causality. Atoms are not moved by forces like love and hate, rather their motion is the consequence of true natural laws.

What started two and a half millennia ago on what is nowadays the west coast of Turkey was nothing less than the beginning of a revolution that continues today. For thousands of years before then, mankind had seen what occurred in the world as coming from some primarily mystic source. Magic and superstition ruled the world.

That changed 2500 years ago on the Ionian coast. The time and place were no accident. In the city states of the Ionian coast democratic values had taken hold. New ideas were accepted easily and could spread quickly. This was due in part to the switch around that time from hieroglyphic symbols to an alphabet. Religion played no or only a subordinate role.

Thus the idea that our world is somehow in the end knowable, and that natural processes can be analyzed with a rational mind, gained ground. "Atomism" was the very beginning of this development. The thread that winds its way through history from the Ionian coast 2500 years ago to this highly scientific and technological present day essentially concerns our knowledge of the building blocks of matter.

Many details about atomic theory as taught in antiquity were preserved for us by a chance event that occurred in Italy in 1417: a manuscript by Lucretius, a Roman poet and philosopher, was discovered. In this script, *De rerum natura*, which is composed in a measured hexameter, Lucretius not only describes the ideas of Leucippus and Democritus but also further develops them. In the work of Lucretius, the atomism of antiquity reaches its highest form. This book was one of the first to be made after the invention of printing. Copies spread all over Europe and have influenced scientists ever since.

In the work of Lucretius, one finds the best and most detailed description of the atomic theory of antiquity, but, in the end, the theory could not prevail against Plato' and Aristotle's system of ideal forms. His work combined scientific questioning and the demystification of nature with a deep respect for nature and its immutable rules.

Had Lucretius's teachings prevailed two millennia ago, the course of world history would have been different. It would have been less marked by religious excesses and the religious wars in Europe and Asia. Alas, the reality was different.

After the collapse of the Roman Empire, the western world sank into intellectual oblivion for more than a millennium, dominated by religious fanaticism and superstition. It was not until the Italian Renaissance that the brilliant intellectual clarity of Greek thought came to wide parts of Europe again, after being lost for more than a thousand years. The scientific epoch commenced then, led by such heroes of the mind as Copernicus, Leonardo da Vinci, Johannes Kepler, and Galileo Galilee.

In the 17th century, the atomism of the philosophers of antiquity was combined with scientific ideals for the first time. At this time, scientists came to the conclusion that chemical elements such as hydrogen, oxygen, and copper were composed of similar atoms. Isaac Newton, the inventor of mechanics and thereby also the inventor of theoretical physics, was even of the opinion that the consistency of materials, the hardness of a metal for example, was somehow linked to the forces between atoms.

In the second half of the 19th century, atomism was applied with success to the field of chemistry. Chemists found that chemical reactions were best understood if one assumed the substances involved to be made up of small, indivisible building blocks, or atoms. A chemical element, hydrogen say, was thought to consist of a single type of atom. And chemical methods made it possible to determine the approximate size of such an atom: 10^{-8} cm. One billion hydrogen atoms stacked on top of one another would reach a height of around 10 cm.

Nowadays, we know of 110 different elements, that is to say, 110 different kinds of atom. This fact would have posed a serious problem for the ancient Greek philosophers: they would certainly never have considered it possible that more than 100 different atoms exist. For the first time, there were doubts about whether the atom was truly an indivisible entity. In the end, it was physicists, not chemists, who realized, at the start of the 20th century, that atoms are not indivisible in the sense understood by the ancient Greeks. They discovered that atoms are made up of smaller components: electrons, the particles that make up the atomic shells, and the atomic nucleus, which constitutes the greatest part of the mass of an atom. Atoms became complicated systems.

In the 1920s and 1930s, atomic physics had its great breakthrough. With the help of the newly developed field of quantum mechanics, it was possible for the first time to understand atoms, and hence the structure of atomic matter, both quantitatively and qualitatively on the basis of a small number of principles. Most of the problems that physicists and chemists had grappled with in the previous century could now be solved elegantly.

Physicists then applied the same principles to the atomic nucleus, in the hope that they would quickly reach a similar, more profound understanding of the nucleus. But this hope was not to be fulfilled. They discovered that the atomic nucleus, far from being an indivisible object, is made up of nuclear particles called protons and neutrons. This knowledge, however, hardly helped to reveal very much about the properties of the atomic nucleus itself.

Soon it was observed that when particles are collided at very high energy, new particles are created. Nobody had anticipated this. Einstein's now famous equation $E = mc^2$, which states that matter and energy are interchangeable, was seen in full effect. A whole zoo of new particles was discovered. Some physicists despaired at this confusing diversity and compared subnuclear physics to botany.

Finally, between 1960 and 1980, a breakthrough was achieved that brought order to the chaotic world of subnuclear physics. The 20th century will enter into history as the epoch in which the substructure of matter was largely elucidated. Today we know that normal matter is made up of quarks, which are the building blocks of the atomic nucleus, and electrons. In the 1970s, a clear picture of the microstructure of matter finally evolved, often prosaically named the Standard Model of particle physics. This model also describes the fundamental interactions qualitatively and quantitatively in a simple form. The interactions are the chromodynamic force, which acts between quarks, and the electroweak force, which acts between quarks and leptons, such as electrons.

The Standard Model is, however, far more than a theoretical model of the elementary particles and their interactions. Its claim to fame is that of a complete unified theory for all the observed phenomena associated with elementary particles. For specialists, the whole theory can be reduced to a couple of lines. This makes it something like the *Weltformel* that theoretical physicists such as Albert Einstein and Werner Heisenberg had unsuccessfully looked for in the past.

Could this theory prove to be a last and thereby final truth? Are electrons and quarks indeed nature's elementary objects, meaning that physicists have finally found the atoms of Democritus and Lucretius? The answer to this question remains undecided. The Standard Model has a number of unsatisfactory characteristics, so many physicists today assume that it is merely an approximation, albeit a very well-functioning approximation, to a more comprehensive theory. If so, physicists should soon find evidence in their experiments for phenomena not explained by the Standard Model, perhaps even evidence for a new substructure of matter.

Electrons and quarks are not simple building blocks that can be combined at will. They are subject to forces such as electromagnetic forces which are, in turn, transmitted by small particles. That is why in particle physics we should avoid speaking of fundamental forces acting on particles, rather we should speak of interactions between particles.

It turns out that interactions in the Standard Model are governed by very specific laws that are based on symmetry. Symmetry in nature and elementary particles are closely connected. Plato had already referred to such a connection in antiquity. Werner Heisenberg, one of the founders of quantum mechanics and one of the most important physicists of the 20th century, had the following to say: "For Plato is the elementary particle not an unchangeable and indivisible object. The elementary particle is reduced to mathematics. The roots of the observed phenomena are not matter, but the mathematical law and the underlying symmetry."

Are ideas finally more important than matter? Or does the distinction between matter and ideas disappear as we attempt to describe the limits of particle physics? Until today, the answer to this question is undecided; it is even unclear whether this is a valid question at all.

Our lives are full of change. Everything is in flux, nothing stays as it was. But this is not the whole truth. There is a continuity in the world that allows us to predict objects and occurrences. We scientists discover that some things stay the same, the laws of nature, for instance. These laws, however, depend on strange numbers that we call the constants of nature. Experiments allow us to determine these numbers with ever increasing precision. But the more precise our answer, the stranger these numbers seem.

The constants of nature reflect a profound knowledge of the universe. They characterize our universe. The fact that they exist tells us that, in other regions of our universe, laws similar to our own apply. But other universes could exist and could have different natural constants. Some physicists don't even speak of "the" universe anymore, because it may not exist in such a form. They speak of a whole collection of universes, called the multiverse.

At the same time, these natural constants stand for our knowledge of the universe and they stand for our ignorance. We do not know how their values come to be. Introducing values that cannot be derived but which must be fixed by experimentation is not satisfactory. Scientists cannot be reconciled to this situation.

Are the values of the natural constants a consequence of some hitherto unknown natural laws, or are they merely a random product of the Big Bang? We do not know. To date, no one has been able to explain the value of any of the natural constants. This is one of the mysteries of science, perhaps the biggest mystery in our world.

Life in the universe is only possible for certain very special values of these natural constants. Why do the natural constants take precisely these values here in our universe? We do not know. What is clear, however, is that if they had other values, life could not exist and there would be no one around to ask the question.

The problem of the natural constants arose due to the increased precision in determining their values that particle physicists have been capable of since the 1970s. That is why the discussions in this book center around particle physics. Quantum mechanics is also discussed along with questions concerning the Big Bang and astrophysics.

It was the Scottish physicist James Clerk Maxwell who first suggested basing certain standards on microscopic objects that are to be found everywhere, such as molecules. Up to that point, standards had been set using objects specially designed for the purpose. Maxwell was impressed by the fact that hydrogen molecules, for example, are the same everywhere, unlike large bodies which have their own particularities. As president of the British Association for the Advancement of Science, a society formed in the 19th century and modeled on the *Gesellschaft Deutscher Naturforscher und Ärzte* (a society of German scientists and medical doctors), Maxwell wrote: "If, then, we wish to obtain standards of length, time, and mass which shall be absolutely permanent, we must seek them not in the dimensions, or the motion, or the mass of our planet, but in the wavelength, the period of vibration, and the absolute mass of these imperishable and unalterable and perfectly similar molecules."

And indeed these ideas of Maxwell are now followed. Our measure of length, for instance, is set by the wavelength of light emitted by atoms of krypton-86. Time is measured by caesium clocks using an atomic transition of caesium.

A large portion of this book deals with questions related to the natural constants that have been introduced in today's Standard Model, or rather, that have had to be introduced in this model. These constants define, to a large degree, the structure of our world. This gives me the opportunity to report on some questions on which I worked with Murray Gell-Mann at the California Institute of Technology in the '70s.

Towards the end of the book, a different question is addressed: are the natural constants really constant? Or are they, however minimally, time dependent? By examining the light emitted by distant quasars, which takes billions of years to reach us, it is possible to study the natural constants in the past. People have found a small yet measurable time dependency of the fine structure constant. Should these measurements truly be correct, the consequences are not yet foreseen. The natural constants could also be minimally dependent on the location in space. That means that the natural constants could have different values in different parts of the universe.

The natural constants confront us with one of the most profound riddles of our universe. Where do they come from? Are they really absolutely constant? Are they the same everywhere? Are they dependent on one another? At this point in time, we can give no definitive answers to these questions. Albert Einstein believed the natural constants to be fixed by the interactions, thus excluding any freedom. To date, however, we have not seen any way of verifying this. Presumably there is freedom in the choice of the constants. It may take hundreds of years to be sure.

At the beginning of the new millennium, particle physics is faced with new challenges, and in all probability physics stands at

Murray Gell-Mann (right) and the author Harald Fritzsch (left) in Berlin in 1995.

the threshold of new and important discoveries. In the 20th century, physicists delved ever deeper into the inner workings of matter. New, previously unknown worlds were accessed, new horizons opened up. The structure of the microphysical world became visible. This complex world can be described by a surprisingly simple theory, a theory that can be formulated mathematically.

The questions to be answered become ever more fundamental. Where does matter come from? What happens to it in the distant future? Where do the natural constants come from? Particle physics remains a great adventure. When a new experiment begins, usually after years of preparation, the physicists involved start on a journey into *terra incognita*.

The abyss between the world of particle physics and the everyday world has become huge. This book aims to reduce this abyss. Why should one set oneself the goal of exploring particles and phenomena that have practically nothing to do with our everyday life? The reasons are the same as those that drive scientists to explore outer space, to push ever further into the depths of our oceans, or to overcome other frontiers. As with all fundamental research, particle physics is part of our culture, part of our effort to gain a rational understanding of the cosmic order.

Fundamental research in physics is, to a large extent, particle physics. The considerable investments made in this field have played a big role in the development of the open and enlightened societies that we find in most parts of the world today. The fascinating insights into the world of the microcosmos that particle physics has allowed us can be counted as one of the lasting achievements of the last century.

One of the essential goals of this book is to inform the general public, not primarily physicists, about the problem of the natural constants. To this end, I use a format that has been used with success before: an imagined dialogue between real and fictitious persons. In this book the persons are Isaac Newton, Albert Einstein, and a modern-day physicist named Adrian Haller, who comes from the University of Bern and is serving as a guest professor at the California Institute of Technology (Caltech) in Pasadena.

The dialogue form has been known since Plato's dialogues in ancient Greece. The famous dialogue *Timaeus*, for example, features the three figures of Critias, Socrates, and Timaeus. Another example is Galileo Galilee's dialogues published in the famous *Discorsi* from the Middle Ages. The format is useful because the reader is often confronted with questions that he may have wanted to ask, and these questions are then answered. The discussions in this book take place in California, in places where I was active years ago.

I wish the reader pleasure, new insights and success in understanding the problem of the fundamental constants.

CONTENTS

THE CONSTANTS OF NATURE

Professor Adrian Haller of the University of Bern sat in a Lufthansa plane en route to Los Angeles. The plane had started from Frankfurt about three hours earlier. He glanced down at the ocean occasionally, concentrating on preparing the lecture that he was soon to give at the California Institute of Technology in Pasadena. This was to take place at the very beginning of his extended stay as a guest professor.

The plane was now over the Atlantic to the west of Ireland. Haller looked down at the water, lost in thought. He felt tired and put down the overhead transparencies for his lecture...

The taxi from Los Angeles airport passed through the city center onto the Pasadena highway, reaching Pasadena at the foot of the San Gabriel mountains in around 40 minutes. The car drove up California Avenue till it reached Hill Street, then turned left on its way to the Athenaeum of the California Institute of Technology. The Athenaeum was the institute's guest house and it had seen many illustrious guests, including Albert Einstein in the '20s. Having received his room key from the receptionist, Haller climbed the stairs and went to his room. It was not long, however, before he picked up the phone to find out whether Newton or Einstein had arrived.

"Of course. They have been here since yesterday and are waiting for you," said the receptionist. "Newton's room number is 119 and Einstein has 137."

A smile played over Haller's lips when he heard Einstein's room number. "Why does Einstein have 137? Did he insist on it?"

"Yes, that's right, he asked specifically for room 137. And since it was available, I gave it to him. I couldn't figure out why he wanted it so badly. Is there something wrong?"

"Everything is fine," Haller reassured the receptionist. "It's just that the number 137 means something special to a physicist. Every physicist knows that, Einstein too, of course."

He made his way to room 119, but found no one there. Finally he knocked on the door to room 137.

"Come in if it's a physicist," called out a voice easily recognizable as Einstein's. Haller laughed and entered. Einstein and Newton were sitting on the couch.

"Hello Professor Haller, we have been waiting for you," said Einstein. "I hope that you had a good trip from old Europe to the New World. Welcome to the country of the sun, California. Have a seat."

Haller: Yes, thank you, the trip was very pleasant. And now here I am, ready for our next discussion round. But let's start it off slow. It was a long flight and I have only just arrived. I'm a little tired.

Newton: It's good to be united again. Einstein and I have been considering what our topic should be for the next few days, but we haven't been able to settle on something specific. We didn't want to preempt you.

Haller: I do have a proposal. It is a topic that is hard to digest because nobody has any concrete idea how to solve the problem. I also have to give a lecture about it at Caltech soon and could use some inspiration.

Einstein: That doesn't sound too bad. Maybe we can contribute something.

Haller: Well, I don't know. The problems are so immense that I have my doubts. But you never know what will come out of the process when Einstein and Newton sit down together to think. I suggest that we take on the topic of the "natural constants". In view of your room number, Mr. Einstein, that seems an obvious choice.

Arnold Sommerfeld (left), who introduced the constant α to physics, talking to Niels Bohr (right). Credit: Niels Bohr Archive, Copenhagen.

Einstein: Ha ha! When I got here, room 137 happened to be vacant, so I took it. Indeed, the number 137 really is very impressive. My friend Arnold Sommerfeld in Munich would have had a laugh at my room choice. He probably would have liked to take the room himself. I am sure that Sommerfeld would have loved to live in room 137. Knowing him, he probably would have been willing to pay twice the price for it.

Wolfgang Pauli (second from right). Credit: CERN, Pauli Archive.

Haller: When your great colleague Wolfgang Pauli died — far too early, not even aged 60 — he passed away in room number 137 of the hospital in Zurich. Most probably it was just coincidence, but one can't exclude that it was Pauli's wish to spend his last days in that room.

Einstein: I would assume the latter to have been the case. Pauli never shied away from doing crazy things. Even his death was no exception. He was just a little nutty.

Newton: But now let us return to the natural constants. In my scientific work, I introduced only one natural constant, the constant of gravitation. Are there other important constants besides this one and Sommerfeld's fine structure constant? I admit that this topic has always fascinated me. Who fixed the value of the constant of gravitation? I am sure there is a lot to say about that. I, for one, don't know the answer.

Haller: Who fixed the value of your constant? The devil himself probably doesn't know the answer. Your constant of gravitation still confronts us with riddles. In the meantime, we have ended up with a whole set of constants, far too many, alas, which all have to be fixed experimentally. Today we distinguish more than 20 different fundamental constants, and all of them must be determined by experiment.

Einstein: My God, more than 20? What then is the meaning of fundamental here? This is more like a continuum of constants.

Haller: I can see already that we should conduct our discussion systematically. I suggest that we begin by studying the fundamental interactions, and that we then move on to the fundamental constants. These two topics are closely related anyway.

First of all, I would like to mention that you, Mr. Einstein, have contributed more to forming our modern view of nature than any other physicist. Although I suppose it doesn't mean much to you anymore, it was you who gave us the correct perspective of the quantum nature of the microscopic world, not to mention the new view we have had to take of the speed of light, c. The value of c is much more than simply the speed of light, it is the fundamental constant of the theory of relativity. Even if there were no light, the constant c would be of utmost importance. And it should not be forgotten that you developed a theory of gravitation that replaced the theory Isaac Newton had given us 250 years earlier, and that linked the gravity force to the curvature of space and time.

But back to the speed of light. Weren't you always fascinated by the idea that things should not change when the observer is in motion? This even applies to the speed of light, which does not vary whatever the state of motion of the observer.

Einstein: Enough of that. That is too much praise for me. Please refrain from such excess in future. It's not praise that is important but criticism, and I don't mind if the criticism is hard, I can take a lot of it. I'm sure it's the same for Newton. What counts is physics, the truth.

In my view, there should be no dimensionless constants that could, from a logical point of view, have completely different values. But I

Isaac Newton. Credit: © National Portrait Gallery, London.

would assume that there are not many people who would agree with me on this point, including you. What really interests me is whether God could have created the world in a different form. That is to say, is there any freedom given the necessity of logical simplicity?

Haller: I'll agree with you on the last part. But I truly cannot share your opinion concerning the constants.

Newton: In any case, why don't we leave these questions about basic principles aside since they don't get us anywhere. We're not philosophers, after all. Einstein always wants to know the whole story, like a real philosopher. I, on the other hand, am quite satisfied with certain approximations.

Let's start right away with our discussion about natural constants, even though you, Mr. Haller, may not really be in the mood as you are still tired from your long flight. What fundamental interactions exist that I haven't heard of yet?

THE FUNDAMENTAL CONSTANTS

Max Planck (left) and Albert Einstein (right) in Berlin.

Haller: Not so hasty, Mr. Newton. As you know, I am a particle physicist. And it was particle physics, or more precisely particle physics since around 1970 that gave us these constants. But before I venture into the realm of particle physics, let me say something about gravity, electricity, and magnetism.

First of all, let's turn to gravitation. Max Planck, who founded quantum theory at the beginning of the 20th century, disliked units such as the kilogram or meter or second, units that are somehow arbitrarily chosen. In their place, he wanted to use units that were as fundamental as possible. He suggested combining the constants of nature in such a way as to arrive at these fundamental units. He used the constant of gravitation, that's your constant Mr. Newton; the speed of light, your constant Mr. Einstein; and the constant h, which he introduced himself. This is the constant of quantum theory and is now called Planck's constant.

In this way, Planck introduced a unit of mass that was named after him, the Planck mass:

$$m_{pl} = \sqrt{hc/G} = 5.56 \times 10^{-5} \text{ g}.$$

The Planck mass unites Einstein, Newton, and Planck. One can also introduce a smallest length, the Planck length, and a smallest time, the Planck time:

$$l_{pl} = (Gh/c^3)^{1/2} = 4.13 \times 10^{-33} \text{ cm},$$

$$t_{pl} = (Gh/c^5)^{1/2} = 1.38 \times 10^{-43} \text{ s}.$$

Einstein: It is especially remarkable that the unit of mass is small, but not really very small. One can compare it to the mass of a bacterium. Planck's length and time units, however, are fantastically small.

Haller: Yes, they are pretty small. But what Planck pointed out is that other observers in the universe would introduce the same units regardless of their location, even if they were near Sirius, say. No matter how small Planck's units may be, the important thing is that they are based completely on fundamental constants and do not contain any arbitrary units such as kilograms or meters, which depend on our particular conventions.

But now to electricity. As we know, electrically charged particles, electrons for example, are deflected by electric and magnetic fields. The deflection is proportional to the charge of the particle. A strange fact is that all the charges that occur in nature are either the same as that of an electron, or the same as that of an electron but with the opposite sign, or a multiple of one of these. The charge of a proton, for example, is the same as that of an electron with the sign reversed.

Newton: Wait a minute, this is strange. I heard that electrons are supposed to be truly elementary while protons are made up of three quarks. Then how is a connection between these charges possible? The charges of quarks are also peculiar: $-\frac{2}{3}$ or $\frac{1}{3}$ of the charge of the electron. Why?

Haller: Today we know that there are probably strict relations between the charge of an electron and the charge of a quark, and these relations mean that the charge of a proton is the same as that of an electron with the sign reversed. We will return to this later.

But back to the charge of an electron. Taking this charge, dividing it by the speed of light c and the constant h of quantum theory (introduced by Planck), and finally multiplying the result by 4π, gives a dimensionless number called the fine structure constant, known as α.

Einstein: I remember it well. After all, it was my friend Arnold Sommerfeld who introduced the constant α in Munich in 1916. We were all perplexed by the strange value of this constant: it is almost exactly $1/137$, the inverse of my room number here! It turns out that this number plays an important role in atomic physics.

Haller: My great colleague Richard Feynman, whom I worked with at Caltech in the '70s, once said to me as we were having lunch downstairs in the Athenaeum: "The number 137 is one of the greatest damn mysteries of physics. A magical number that comes to us without anybody understanding it. One could say that it was written by God to make a fool of us." One day in 1975 I was once again, as so often in those days, having lunch with Richard Feynman. We spoke about the strange number 137. Suddenly Feynman grinned and said: "You know, every theoretician should write: '137 — how little we know' on their office blackboard. That would teach them humility."

I laughed and said that I would write that on my blackboard immediately. However, on my way back from lunch, I first passed by Feynman's office. There were equations on his board, but of course there was no sign of the number 137. So I took some chalk and wrote:

137 — how little Feynman knows

An hour later, Feynman came by to thank me for the inscription. Not seeing one on my blackboard, he wrote:

137 — how little Haller knows
Signed: Richard Feynman

Even though Gell-Mann was constantly urging me to wipe away "Feynman's nonsense", I left this inscription on my board until I left Caltech in March 1976. It was too bad that I didn't take my blackboard with me to Geneva; I really should have done so. Later Feynman would often visit me in Geneva, but every time I forgot to ask him to write that little sentence on my board again.

Feynman died in February 1988 of a nasty disease that he had been suffering from since the end of the '70s, and so I was left without an inscription on my board by this great physicist. I do, however, have a personal dedication from Feynman in a book that he wrote and gave to me as a present. It was the first copy that he had received from the publisher and he gave it to me because I was leaving for Europe that very day. I treasure this book about the life of Feynman as if it were a holy relic.

The American physicist Richard Feynman (1918–1988) was one of the pioneers of quantum electrodynamics, which describes electromagnetic quantum processes. He worked at Caltech in Pasadena from the '50s.

Einstein: But back to electrical forces. There have, in the past, been a number of attempts to define α mathematically. In Germany, Werner Heisenberg proposed:

$$1/\alpha = 2^4 3^3 / \pi.$$

But this doesn't work that well because it gives 137.51. Heisenberg would probably not be proud of this publication from the '30s anymore.

Haller: In 1971, I was working at CERN. One day, a mathematician from Zurich named Mr. Wyler gave a lecture and derived the following:

$$1/\alpha = (8\pi^4 / 9)(2^4 5! / \pi^5)^{1/4} = 137.036082....$$

This expression works quite well — to about one part in a million. Moreover, it is not an arbitrary construct, but rather the ratio of two group spaces in mathematics. This I mention only in passing because Wyler himself did not have the slightest idea why just such a ratio should be relevant for the electromagnetic interaction. Still, someone might find a use for this formula.

There is another number that we should familiarize ourselves with. The gravitational constant introduced by our friend Newton is today known quite exactly:

$$G = 6.67259 \times 10^{-11} \text{ m}^3\text{s}^{-2}\text{kg}^{-1}.$$

This constant has strange dimensions, but we can avoid them by multiplying G by the square of the mass of a proton and dividing by hc.

$$\alpha_G = Gm_p^2 / hc \approx 10^{-38}.$$

The dimensionless number that we arrive at is extremely small. This of course reflects the fact that gravitation is very weak indeed in comparison to the other forces that we observe in nature, such as the electric force.

But let me check the time. It's already eight o'clock: time for dinner. I suggest that we cease our discussion and go directly to the Athenaeum's restaurant.

A few minutes later, they sat down at a table in the Athenaeum and studied the menu. Haller decided for all of them: filet mignon and a bottle of good Californian red wine, Cabernet from the Napa valley, north of San Francisco.

The food was served quickly and the three of them ate the substantial meal with relish. Afterwards, Haller ordered a dessert, just a lemon sorbet — of which he was very fond — and a coffee with milk. But that was all for the day. They retired to their rooms. Haller, who was already very tired, fell asleep immediately.

ELEMENTARY
INTERACTIONS

2

The next morning, the three physicists met in the Athenaeum for breakfast. Haller suggested that they continue their discussion in the Athenaeum's small library, which Einstein knew well, and his proposal was immediately accepted. They went there right after breakfast. Haller first made a quick run back to his room to pick up a picture, which he showed to Einstein and to Newton. Einstein recognized it immediately.

"Yes, this photo shows me here in this room, in 1929, writing my equation of general relativity on the blackboard. Really, that was more of a gag for the journalists who were present than it was meant for the rest of the audience. Robert Millikan, who was at that time president of Caltech, was one of the people there. He later told me that he had actually understood very little of my speech. It wasn't that surprising because my speech was meant for the Caltech physicists. Although Millikan was, of course, a physicist too, he was no longer active and he was not so up to date. He was kept busy enough with administrative work in his role as Caltech president."

Haller: I once had something to do with this photograph. I produced a TV serial called "Mikrocosmos" in 1985 for the WDR in Cologne.

Albert Einstein. Credit: AP Photo.

It was a six-part TV serial about particle physics, meant for the layman rather than for physicists. We took this photograph of you, took out your equation of general relativity, and then I wrote your famous equation

$$E = mc^2$$

in its place on the blackboard.

I also used this edited photograph on the cover of a book that I wrote about the theory of relativity. Since people do not know this photograph of you and the equation, they sometimes ask me where I got it from. I am constantly explaining that it is not a real photograph, but rather a photo-collage.

A while back, I flew into San Francisco, and as I was driving my rental car from the airport to Stanford University, I suddenly saw on a huge screen this picture of Albert Einstein with the formula $E = mc^2$ in my handwriting. It was an ad for some product. I found it strange that my handwriting was there.

Einstein: But let's discuss the fundamental interactions. This is largely new territory for Newton and me, so I suggest that we sit down and that you give us a lecture.

Haller: That's a good idea. I will begin immediately and then, if everything goes well, we may finish today.

If two bodies influence each other, it means that there must be some form of contact between them, and this can be realized in different ways. Mostly there is a force between the two bodies, such as when the Earth pulls on an apple hanging from a tree until the apple finally falls off the tree. As it turns out, it is not only the massive body of the Earth that exerts a force, namely the gravitational force, on the apple; the apple also attracts the Earth. We are dealing here with a reciprocal phenomenon, and this is always the case when there are forces acting between bodies. That is why in physics we often speak of interactions rather than forces. In the example I just gave, the interaction would be gravitation.

But now it gets more complicated. Gravitation is not the only interaction that we observe as a macroscopic phenomenon in nature, even though it may be the one that occurs most frequently and that is often taken advantage of, for example when skiing. We can also observe electric forces in every day life, and it is common knowledge that bodies with the same charge repel each other, while those with opposite charges attract each other. Magnetic forces also turn up frequently e.g., when a compass needle orients itself, or when magnetic resonance tomography is used in a check-up at the hospital.

Until the beginning of the 19th century scientists differentiated between electric interactions and magnetic interactions; the two were seen as separate phenomena. This all changed when it was discovered that electric currents or moving charged bodies induce magnetic forces.

Also important was the inverse of this effect, namely the generation of electric currents by rapidly changing magnetic forces. Today we use this effect to generate electricity in power stations.

Einstein: And remember what the English physicist Michael Faraday, anticipating the future, once said? The British Minister of Finance, who was visiting him, asked Faraday whether the electric currents that he was experimenting with were really all that important. Faraday answered: I bet you that one day you will collect taxes on them. And he was right.

Haller: The reciprocal relations observed between electric and magnetic phenomena give us reason to believe that electric and magnetic phenomena are closely related. Michael Faraday introduced the concept of electric fields and magnetic fields. He fundamentally changed the way that scientists think about the generation of forces.

Till then, it had been usual to imagine electric, magnetic and even gravitational forces as phenomena that mysteriously act over some given distance. In this view, the Earth attracts the Moon because the Earth exercises a force on the Moon across a distance of 300,000 km.

Faraday, on the other hand, imagined electrical forces, for example, to be the result of a field generated by a charged body, filling the adjoining space with lines of force. Two electrically charged bodies attract each other because the space between them is occupied by a field — the space is "filled", so to speak, with lines of force. Faraday really believed that these lines of force were there, he could feel them in his fingers. That's typical of an experimental physicist. A theoretician would never have thought of such a thing.

Einstein: Objection! I could have come up with that idea.

Haller: OK, but you are Albert Einstein. As a physicist, you are a generalist rather than strictly a theoretician. Moreover, you are a genius. But let's continue. Electric and magnetic field lines influence each other: this is the way to understand the reciprocal relationship between electric and magnetic phenomena. This is also the reason why, in general, we speak of electromagnetic fields. It later became clear, in the context of the theory of relativity, that electric and magnetic fields must be considered as a unit. So Faraday was right with his idea that electric and magnetic fields imply each other — one cannot exist without the other. A magnetic field that is examined by an observer who is moving very fast will appear as a mixture of an electric field and a magnetic field.

Einstein: Yes, the idea of fields was a great achievement of Faraday, who after all had no inkling of mathematics. Without Faraday, I certainly would never have invented my theory of gravitation, and probably not the theory of relativity.

Haller: The concept of a field has, in the meantime, become one of the most important physical concepts, without which no quantitative description of natural phenomena would be possible. Even so, we are dealing with an abstract concept, one which presents a difficulty for every layman because we cannot detect fields with our senses, or can do so only very indirectly. Humans can feel sound waves, but we do not feel electromagnetic waves except for visible light, which we see with our eyes.

Today the theories we use to describe the behavior of elementary particles are, without exception, field theories. Every physicist or engineer alive today knows that fields are a reality. Fields have energy and momentum, just like material bodies.

Fields are not exclusively connected to objects; they can have a life of their own. The equations that describe the behavior of electromagnetic fields were found by Maxwell in 1861 and were named after him. Besides Newton's laws of mechanics, the Maxwell equations represent the theoretical pillars upon which a large part of modern technology is based. Later Heinrich Hertz in Germany discovered the electromagnetic waves in his laboratory.

Einstein: Clearly Maxwell would have deserved for his achievements a fair remuneration from industry, from Siemens, AEG, etc, mostly from the Germans.

Haller: Sure, many millions of dollars. I would also like to emphasize that the Maxwell equations are relativistic. Maxwell could have found the theory of relativity a long time before you, Mr. Einstein.

Einstein: No doubt. But luckily for me, he didn't. He was probably very close, but he did not notice that his equations already implied the theory of relativity. Maxwell died very young, at the age of 48. If he had lived another 40 years, he might have found the theory of relativity.

Haller: One of the important consequences of the Maxwell equations is that time-dependent variations of electric and magnetic fields spread out in space at the speed of light. That is no coincidence, because light itself is an electromagnetic phenomena. Light waves are electromagnetic waves that our eyes can perceive. They have a wavelength of about 0.001 millimeters.

X-rays are nothing more than electromagnetic waves that have a considerably shorter wavelength than normal light. Radio waves have a longer wavelength: shortwave radio uses wavelengths that are in the range of a couple of tens of meters.

Einstein: Let me mention that electromagnetic waves possess quantum properties. When electromagnetic waves spread out in space at the speed of light, the energy is not propagated continuously, as we would expect from classical wave theory. Rather, it is transmitted in the form of packets of energy called photons. I introduced the concept of a photon in 1905. The Nobel Prize that I received for this discovery I gave to my wife. I would never have received such a prize for the theory of relativity, it was far too speculative and too far away from the real world for the men of the Nobel Prize committee.

Haller: That's for sure. On top of that, the theory of relativity was considered very uncertain. But the Nobel Prize for photons was nevertheless well deserved. It was a good idea.

Now back to Maxwell's equations. The equations of electromagnetism describe the dynamics of photons. But it is important to interpret them in the context of quantum physics. This leads to the theory of quantum electrodynamics, or QED. This theory was developed in the course of the 1930s, sadly not by you, Mr. Einstein.

Einstein: Don't remind me. It was a stupid mistake not to do it. Instead I occupied myself with useless theories.

Haller: Never mind, you achieved enough. Two people who made important contributions were Werner Heisenberg and Wolfgang Pauli. They based their ideas on the work of the English theoretician Paul Dirac, whom we will mention again in another context. It is interesting to note that quantum electrodynamics unites quantum mechanics, field theory, and Einstein's theory of relativity.

In the context of electrodynamics, one finds a new and important interpretation of the electromagnetic forces. Consider the scattering of two electrons, for instance. The two particles move toward each other, pass each other, and then move away from each other. Since each particle is repelled by the other due to the electric force, they both change direction. They are scattered at certain angles, which depend on the details of the particles' trajectories.

How can one picture this scattering from the point of view of quantum electrodynamics, in which the electromagnetic field has quantum properties and its energy is measured in photons?

An electron is surrounded by an electromagnetic field that is described in terms of photons. A fast-moving electron can be pictured as a massive charged body surrounded by a cloud of photons that move through space

Paul Dirac (left) and Werner Heisenberg (right) in Cambridge, England in the 1930s.

with the electron. When two electrons fly past each other, the two clouds mix. Since photons do not come with name-tags identifying them as belonging to one particular electron, they can be exchanged. Photons from one electron can, after the encounter, find themselves part of the cloud of the other electron and vice versa. Since photons carry energy and momentum, this leads to a change in the momentum and hence a change in the flight direction of each of the two electrons, which is interpreted as the action of a force. So the force results from an exchange of photons. That is why we should speak of an interaction here rather than of a force.

In the case of two electrons, we have a repulsive interaction. If we observe the encounter of an electron with a positron (that is the antiparticle to an electron, of which we will speak later), photons are once again exchanged between the clouds, but this time the net result is an attraction.

What is important here is that quantum electrodynamics states that the forces acting between charged particles are themselves transmitted by particles, namely by photons, the same quantum objects that manifest themselves in nature as the particles of light.

There is, however, an important difference between particles of light and the photons that transmit electromagnetic forces. The photons of light are referred to as free or real photons, since they are independent entities, not part of a cloud belonging to some charged particle. The rule for these is that they always have an energy equal to the absolute value of their momentum. For example, a photon with an energy of one mega electron volt (MeV), created in a nuclear reaction, has at the same time a momentum in a certain direction with an absolute value of one MeV.

Einstein: We still haven't told Newton what an eV, or electron volt, is. It is defined as the energy that an electron receives when it crosses a voltage difference of one volt — it is a very small amount of energy.

Haller: Particles with the property I just described have no mass. A particle with a mass m, when at rest, has an energy of $E = mc^2$ and no momentum. This means that the requirement for "masslessness", that energy is proportional to momentum, cannot possibly be fulfilled at rest. But for photons this is not a problem since they are never at rest; they always move at the speed of light.

As a result of the uncertainty principle, which we will soon look at more closely, the photons that are responsible for transmitting energy between two charged particles can have any energy and momentum. They are referred to as virtual particles. The electrical attraction between two oppositely charged bodies comes from the exchange of virtual photons.

The strength of the interaction between two charged bodies depends on the strength of the interaction between the photons and the charged particles. Even though the electrical attraction or repulsion occurs across some distance, the actual interaction takes place between the electron or positron and the photon. We call this a local interaction because the contact between the charged particle and the photon takes place at a point.

Einstein: But now you really must say something about the fine structure constant.

Haller: Yes, I was just getting to that. In Munich in 1916 Arnold Sommerfeld noticed that the strength of the interaction between photons and electrons was described by a number that he called the fine structure constant α, of which we spoke earlier. As the name suggests, this number tells us something about the fine structure of the atomic energy levels. In this number the theory of relativity, quantum theory, and electrodynamics come together.

The constant α is given by the expression $4\pi e^2/hc$, where e is the unit of electric charge. This expression involves e, the electric charge, representing electrodynamics; the constant h from quantum theory; and c, the fundamental constant of the theory of relativity, the speed of light — Maxwell, Einstein, and Planck are unified.

The fine structure constant was the second fundamental constant that entered into physics after the constant of gravity had made its debut, introduced by you, Mr. Newton. But unlike the constant of gravity, the fine structure constant is a pure number, meaning that it has no dimensions such as meters or seconds. It must be deduced experimentally. Today its value is known very precisely, but for our purposes a precision of one part in a million will suffice: we take $\alpha = 1/137.036$. So we are talking about a small number, a little less than 0.01. The reciprocal value of α is almost a whole number, namely 137, which is a prime number.

This number has been the subject of much speculation since its introduction. The reason for all the speculation is that α describes the strength of interactions, making it of fundamental importance to all of science and technology. Because the structures of atoms and of molecules depend decisively on this constant, many things in our daily life would be different if α had a different value from the one observed.

If α were just a little smaller than observed, then many complex molecules, for example, would not exist as stable systems, and this would have unforeseeable consequences for biology. That should make it clear that it would be a considerable advance in the understanding of fundamental interactions to be able to calculate α theoretically, but I suspect that such a calculation is not possible.

Einstein: I disagree. I do think that it will eventually be possible to calculate α.

Haller: I do not wish to argue with you about it now. Let's continue. As I mentioned before, in QED we have a local interaction between a charged particle and a photon. When two electrons repel each other, the electromagnetic interaction actually takes place twice, once when the virtual photon is sent out at a point, known as the vertex, and once when the photon is absorbed at another vertex.

Processes of this nature are described by diagrams named after Richard Feynman here at Caltech. He invented these diagrams.

He had a small van with a diagram on each side. He told me a story once about stopping to get gas at a gas station in Arizona. The attendant came to him and asked him why he had Feynman diagrams on his van, to which he proudly answered: "I am Feynman." The attendant was very moved and refused to take any money for the gasoline "from a Feynman".

By the way, it might seem somewhat strange that the charge of the most important electrical particle, the electron, is negative. It would be better if this charge were positive. But the expressions "positive" and "negative" go back to Benjamin Franklin; he defined them that way and no one can argue with Franklin.

The fine structure constant is quite a small number. This has remarkable consequences for the quantitative descriptions of processes in

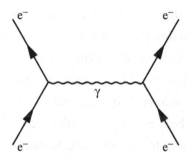

The repulsion between two electrons is described in quantum electrodynamics as the result of an exchange of a virtual photon between the two electrons. This is shown by a Feynman diagram. The local interaction takes place at two points known as vertices. By moving from one point to the other, the virtual photon transmits the force.

quantum electrodynamics. The electromagnetic interaction is quite weak. When an electron interacts with matter, by colliding with an atom, for example, the result is mostly not very dramatic. The electron is usually only slightly deflected.

It is possible to calculate the quantum corrections to the basic processes of QED. This procedure is known as perturbation theory. Here we deal with processes of so called higher order, when two photons instead of one are exchanged, for example. In this case, the basic electromagnetic interaction must take place four times. That means that the strength of the process is proportional to the fourth power of e, or α^2.

Since α is around 0.0073, the effect we are talking about is of the magnitude of α^2, or 0.00005. Other quantum corrections are proportional to the third power of α, or 0.0000004, etc. In spite of the smallness of these quantum corrections, they can be determined experimentally as well as theoretically, and today the agreement between experiment and theory is very impressive. They match to one part in a billion.

Einstein: It can surely be said that QED is the most successful theory of all so far. It's really too bad that I didn't have more to do with it.

Haller: I would agree with you. QED is important as an example of a successful quantum field theory. All theories in particle physics today are constructed using QED as a model.

For the theory of QED it is important that we consider, in addition to electrons and photons, antiparticles called positrons. Without the positrons, QED could not be formulated as a field theory.

To be able to successfully describe the electromagnetic interactions in quantum theory, it is absolutely necessary for antiparticles to exist. The English physicist Paul Dirac realized this even before the positron was discovered in 1932. Towards the end of the '20s, he attempted to connect the newly found laws of quantum theory with the theory of relativity. In the process, he discovered a formula that was then named after him: the Dirac equation. This equation was extremely important for the further development of quantum electrodynamics and particle physics. We will return to the Dirac equation later. In any case, the only way to deal with this equation consistently is if, alongside electrons, there also exist antiparticles with a positive charge, the positrons.

It is assumed in the theory that these particles — electrons, positrons and photons — have no inner structure. According to the theory, an electron is a point mass that possesses an electromagnetic interaction. But the situation is not quite so simple. A consequence of quantum theory is that empty space is not really empty. Rather it is filled with virtual electrons and positrons, which exist only in very small intervals of time and space.

Einstein: That means that an electron is actually a complex structure that is surrounded by a cloud of electron–positron pairs.

Haller: Yes, an electron has a cloud, but we can calculate the effects of this cloud exactly. In a vacuum, pairs are constantly being created and destroyed. A vacuum has nothing like the calm structure it has in classical physics. The smaller the distance at which one chooses to examine the vacuum, the more violent the processes of the virtual particles become.

A vacuum at very small distances looks more like a bubbling cauldron than an empty space.

You can compare this to an ocean. If you fly over an ocean at a great height, you see a flat and seemingly endless surface of water under you, a good approximation to a perfect two-dimensional space. This changes gradually as the airplane looses height. At first one observes light disturbances on the surface of the water; they later turn out to be considerable waves. For empty space, or the vacuum, it is the same thing. From a large distance, the vacuum really is the empty, calm space that we can perceive macroscopically with our senses. But in small intervals of time and space, it is not like this at all. Electron–positron pairs and photons are constantly being created and destroyed.

Newton: So on a microscopic level, there is constant hectic activity, a dance of virtual particles. But we, from our macroscopic perspective, do not notice this because the effects cancel each other.

Haller: These processes do not have direct macroscopic consequences, but they do influence the space surrounding an electron. This can be pictured as follows: if we were to put an electron at a certain point in space, it would repel the virtual electrons while attracting the virtual positrons. So, the vacuum will become polarized. In the vicinity of the electron there will be a surplus of virtual positrons. As a result, some of the electron's charge is blocked by this cloud of positrons.

Newton: So that means that if we examine an electron from the outside, we don't see a point-like electron, but an electron together with its cloud of virtual particles.

Haller: Yes, we call this a physical electron, in contrast to an electron without its polarized cloud, which we refer to as a naked electron. A naked electron must have a greater charge than a physical electron. A physical electron is an actual particle while a naked electron is not real; it is just a theoretical construct.

So as a consequence of quantum theory, a point-like electron is not really so point-like. From a relatively large distance it may look like a point, but if we are within a hundredth of the diameter of an atom, we begin to see the effects of the polarized vacuum. In the context of quantum electrodynamics these effects are very noticeable.

If we calculate what the charge of a naked electron must be, compared to the measured charge of a physical electron, we arrive at a nonsensical answer: it is infinite. This is no great surprise because the naked electron, as mentioned before, is a theoretical construct.

This is not the only unpleasant surprise that occurs in the framework of the theory. We find something similar if we examine the mass of an electron. The equivalence relation between energy and mass means that the electric field of an electron will contribute to its mass, since the electric field means there is a density of energy in space.

If we calculate the mass contributed by the field, we once again obtain a nonsensical infinite answer. This is also completely comprehensible since, in the framework of the theory, we take the electron to be point-like, having no inner structure. Accordingly, at small distances the field will be very strong and a quantitative analysis then predicts that the contribution of the field to the mass is infinite. So we see that the assumption that the electron is point-like leads to nonsensical infinities.

It is possible that the electron does posses an inner structure, which is only noticeable at very small distances, say 10^{-18} cm. That would mean that the electron, rather than being infinitely small, has a very small but finite radius. One can then easily convince oneself that infinities do not occur anymore. The radius of the electron appears in their place in the calculations.

Despite great efforts, it has not yet been possible to find experimental evidence for a substructure of the electron. Instead, all we can say is that if the electron has a nonzero, that is not infinitely small, radius, it must be under the threshold of one hundredth the size of the nuclear particles.

Considering the nonsense that quantum electrodynamics produces, it is fair to ask whether it can be called a successful theory at all. The answer is nevertheless positive. The infinite charge of the naked electron is not really a big problem, because the naked electron does not exist as an actual particle. It is merely a theoretical construct, a product of our imagination that does not come into play until we separate the electron in an artificial way into a core, the naked electron, and the surrounding shell made up of virtual particles. Nature does not make this separation.

We can simply "absorb" the infinity by setting the charge of the physical electron equal to the experimental value and ignoring the charge of the naked electron. In the same way, we set the mass of the electron equal to its experimentally found mass, and ignore the fact that the calculation gives a nonsensical answer. This procedure is called renormalization. It can be shown that this allows a logically consistent description of quantum electrodynamics. The formal infinities do not occur when we replace them with physical values.

Richard Feynman, one of the inventors of this method, once spoke at a conference in the United States and showed that the infinities cancel out if one allows only physically measurable values. His skeptical colleague Robert Oppenheimer, the man who directed the American atomic bomb project during the Second World War, said: "Mr. Feynman, one should not necessarily deduce from the fact that something is infinite, that it is zero." The irony in these words was not so fitting after all. At least from today's perspective, the pragmatism inherent in the principle of renormalization has proven to be thoroughly justified. What's important is that this way of describing the quantum nature of electrodynamic processes has produced a number of notable successes.

Soon after he found his equation, Dirac could prove that electrons must have magnetic properties, described by a certain magnetic moment. In fact, the magnetic moment that Dirac calculated fit the experimental results. It was only decades later that a small divergence was found between experiment and Dirac's value for the magnetic moment, and the difference was very small. The measured moment was only about 0.1% larger than Dirac's value. This presented a challenge to the theory, which was then solved.

It was possible to show that, in the context of QED, the small divergence was an effect of renormalization, that is to say an effect of the virtual particles surrounding the electron. The infinities disappear in the calculation of the magnetic moment and our result is so simple that we can even state it here: the magnetic moment is greater by an amount equal to $\alpha/2\pi$, which is numerically almost 0.1%.

Today the magnetic moment is far more accurately known, and to gain a theoretical understanding of it, one must perform very carefully the calculation of the quantum processes associated with virtual particles. In any case, the agreement between the theory and experiments is breathtaking. This underlines the fact that quantum electrodynamics really is a correct theoretical description of microphysical processes — a triumph of theory.

Einstein: Yes, that does sound pretty good. I must confess that at the beginning of the 1950s I didn't think much of it. Even though I read Feynman's papers, I didn't understand them. Feynman was certainly a good speaker, but his written papers were really quite unreadable.

Haller: Yes, I agree. Feynman's early works, especially, are hard to read. As a physicist, Feyman was an intuitive thinker, and he didn't take the written presentation of his ideas so seriously. Anyone who couldn't catch exactly the drift of his intuition had problems reading his publications, and that was everybody with one exception: Feynman himself.

But now I suggest that it's time for a coffee break. Why don't we go to the restaurant?

With that, the discussion was interrupted. The three physicists went to the Athenaeum's cafeteria.

ELEMENTARY
INTERACTIONS AND
QUANTUM THEORY

After their coffee break the three physicists resumed their discussion in the library.

Haller: I would like to mention another quantum effect that is related to the electric charge. As I explained earlier, an electron's charge is shielded by a cloud of virtual particles that surround the electron. If one were to remove a part of the cloud, then this shielding effect would be a little smaller, and the electric charge of the particle would be greater than it was before. This means that the effective value of the fine structure constant would get a little bigger. This constant is not actually a constant, but a function of distance.

It is not possible to remove part of the electron cloud. But there are ways to get around this, for example by considering a collision of an electron with another electron or a positron at high energy. In this case the particles come very close, and so part of the shielding effect of the cloud is neutralized. During the collision the particles behave as if their charges were a little larger than would be measured in atomic physics.

Experiments of this type were conducted in the last decade of the 20th century with the particle accelerators LEP at CERN and SLC in Stanford. The value for α measured in this way proved to be 7% larger

The ring of the LEP collider at CERN lies between Geneva's airport and the Jura mountains.

than the value measured in atomic physics. The difference corresponds exactly to the theoretical expectations. Once again, the agreement between theory and reality could not be better.

Einstein: That is remarkable! It seems to me that the quantum physics works very well.

Haller: Yes, the theory of QED had been introduced by Dirac, Heisenberg and Pauli as a description of atomic physics phenomena but it turned out to be very successful also at very high energies. Even at energies a million times higher than the typical energies of atomic physics the field equations of QED have proved to be valid. This alone justifies calling the creation of QED one of the intellectual masterpieces of the 20th century.

In fact, the theory of QED is more than a theory of electrons and photons. All modern theories of fundamental fields and particles are very similar to QED. We are going to return to this in more detail.

But first I will mention one important property of QED. When the German mathematician Hermann Weyl looked closely at the field equations of QED in the 1930s, he noticed a new symmetry. This symmetry, which had not been noticed by Heisenberg or Pauli, was later named a gauge symmetry.

The quantum field that describes electrons and positrons is a complex field. This means that, at every point in space, the field is described by a complex number. A complex number is composed of two numbers which are usually denoted as a point in the complex plane. It turns out to be especially useful to describe such a number by its distance from the plane's origin and an angle, known as the phase angle.

In the context of QED, the phase angle of the field of the electron is not specified. It can be set arbitrarily, meaning that one can rotate the phase angle without causing any change in the physical properties of the electron. This transformation is called a gauge transformation. But note that when you make such a transformation, it is important to make the same turn at every point. If we turn the phase angle of the field in Paris by 20°, we must also turn it in Berlin by 20°.

Newton: What a strange condition — why should the field in Paris know when the field in Berlin is turned?

Haller: That is exactly the problem. If, though, we were to turn the field differently at different points: a turn in Berlin of 10°, in New York of 37°, and in Tokyo of 73°, for example, then the scheme is not consistent. So it is possible to carry out a global phase transformation, but not a local one. That is why we refer to a global gauge symmetry rather than a local gauge symmetry.

Understandably, Weyl was not satisfied with this property, and he tried changing the equations so that it would be possible to turn the field arbitrarily at any point. For the field of an electron that is not undergoing any interactions this, as I just said, is not possible. But it becomes possible if the electron is involved in interactions with an electromagnetic field, as we observe in nature. The changes that appear in the equations when we turn the phase angle differently at different points are then compensated by changes in the description of the electromagnetic field, i.e., by a change of gauge of the electromagnetic field.

The two gauge changes, the transformation of the electron's field by turning the phase angle and the transformation of the electromagnetic field, must mesh perfectly, like two cogwheels in a car engine. Only then can a force be transmitted. This is a local gauge symmetry. The meshing of the two fields is only possible when the quantum of the electromagnetic field, the photon, has spin 1, as happens to be the case. If photons had spin 0, i.e., no spin, the meshing would not be possible. It just would not work.

Electromagnetic forces are, in this way, a consequence of a local gauge symmetry. An interacting system of electrons, positrons and photons therefore has a very high symmetry. It is simpler than the symmetry that exists when there are no interactions. Furthermore, it is possible to show that the local gauge symmetry requires the photon to be massless. Although it would definitely be possible to change the equations of QED to give the photon a certain mass, we would then have to give up the local gauge symmetry. So masslessness and gauge symmetry are closely connected.

A further consequence of this symmetry is that the electric charge is a strictly conserved quantity: a charge cannot be created or destroyed. If a system has a certain charge, then, assuming that it has no contact with other systems, this charge will not change over the course of time. Charge conservation and local gauge symmetry are closely intertwined.

In summary, QED is a theory of charged particles and photons that is based on local gauge symmetry. For this reason, we also call it a gauge theory, an expression that was not used by Weyl but which came into use 40 years after his work, at the beginning of the '70s.

Einstein: It was not just chance that led Weyl to discover the local gauge symmetries in QED. Weyl was following a theoretical path that I had been down before. I published my theory of gravitation — general relativity — in 1916. Gauge transformations are also possible in this theory. They are of a different sort to the transformations in QED, but we don't want to go into detail about the differences here. It is basically about changing the coordinate system. As in QED, gauge transformations in my theory are closely tied to conservation laws. In the case of gravitation, these are the laws of conservation of energy and momentum.

Haller: Yes, and we will soon see that the idea of the local gauge symmetry goes far beyond the framework of QED, and is of fundamental importance for the description of the interactions of elementary particles.

But now back to atomic physics. Towards the end of the 19th century it turned out that atoms were not, after all, indivisible, but could be split relatively easily into their constituents. The size of an atom is given by its cloud of electrons.

The mass of an electron and that of other particles is not usually given in kilograms, but in units of energy, in electron volts. According to the equivalence between mass and energy, expressed by your equation $E = mc^2$, it is possible to give the mass of any object, including that of an electron, in electron volts.

An electron volt, or eV for short, is the energy that an electron acquires when it is accelerated through a voltage of one volt. If we use a 1.5 V battery to charge two metal plates that are facing each other, an electron near the negatively charged plate will be accelerated towards the positively charged plate. When it gets there, it will have an energy of 1.5 eV.

For many aspects of particle physics, the eV is too small a unit. We often use the kilo-electron volt (1 keV = 1000 eV), the mega-electron volt (1 MeV = 1000 keV), the giga-electron volt (1 GeV = 1000 MeV) and the tera-electron volt (1 TeV = 1000 GeV). To put this in perspective, the electrons that make the picture on a TV screen have an energy of a couple of tens of keVs.

The LEP accelerator, which operated at CERN until the end of 2000, was capable of accelerating electrons to an energy of about 110 GeV.

But let's get back to atoms. The mass of an atom is concentrated almost completely in its nucleus. The mass of an electron is only about 1/2000, or more exactly 1/1837.2, of the mass of the lightest atomic nucleus, the nucleus of a hydrogen atom. When the electron was discovered, there was speculation that the substance of an atom consisted of electrons embedded in diffusely distributed, positively charged matter.

Einstein: Even I thought at first that it would be something like this, but the idea was totally wrong.

Haller: Yes, just after the start of the 20th century it was established that the main contribution to the mass of an atom came not from its electrons, but rather from a massive nucleus. The decisive experiment was performed by Ernest Rutherford and his co-workers at the University of Manchester, England, in 1910. They worked with alpha particles, which are particles emitted by a radioactive substance. It later turned out that alpha particles are nothing more than helium nuclei.

The alpha particles were directed through a thin gold film. The whole apparatus was surrounded by screens of zinc sulphide. When an alpha particle hit a zinc sulphide molecule, a flash of light was produced which could be seen with the naked eye in a darkened room. This simple setup made it easy to see if any of the alpha particles had come into close contact with an atom when passing through the film: if they had, the particle would change direction of motion, i.e., it would be deflected. Mostly these deflections were very small, but occasionally there would be a strong deflection of 10° or more.

Rutherford and his co-workers proceeded systematically and checked if any of the alpha particles were flying into the gold film and flying back out afterwards, like a tennis ball hitting a solid wall. Nobody thought that this could happen, but it did. On average, one in every 8000 alpha particles met this fate. The particle would be deflected backwards, much to the surprise of the physicists performing the experiment.

"It was quite the most incredible event that has ever happened to me in my life. It was almost as incredible as if you fired a 15-inch [artillery] shell at a piece of tissue paper and it came back and hit you," Rutherford later said.

Almost two years passed before Rutherford understood the theory of what was going on. There was only one way to explain the strange backwards deflection: practically the whole mass and positive charge of the gold atoms in the film must be concentrated in a very small volume at the center of each atom, forming a nucleus, while the electrons, which contribute very little to the total mass of the atom, buzz around in the whole atom.

Further experiments made it possible to estimate the size of the nucleus. It was found to be about ten thousand times smaller than the atomic radius, so around 10^{-12}–10^{-13} cm. The volume of the nucleus is therefore almost ridiculously small in comparison with the atomic volume — atoms are mostly empty space. When one touches the surface of a diamond, one has the impression that it is a very hard object. It is the interplay of the electric force between the nucleus and the electrons that generates this hardness. Even a diamond is made up mostly of empty space. Alpha particles will pass through a diamond just as easily as they pass through a gold film.

What did Democritus once say? Nothing exists except atoms and empty space. What would he say today if he heard that atoms themselves are mostly empty space?

Rutherford often emphasized that an experimental physicist needs a fair portion of luck to make an important discovery. As it later turned out, this was true for his alpha particle experiment. By chance, Rutherford had used a radioactive source that produced alpha particles with an energy of 5 MeV. This energy was ideal for discovering the atomic nucleus. Had the particles' energy been any higher, the alpha particles would have undergone complicated reactions with the atomic nuclei. This would have made it impossible to interpret the results in any simple way. On the other hand, had the energy been any lower, the backwards deflection would not have been observed at all.

Einstein: Yes, Rutherford really was very lucky. Experimental physicists always need luck, and the good ones get it, at least sometimes.

Haller: Rutherford was able to show that the deflections of the alpha particles followed a simple force law. Alpha particles and atomic nuclei are both positive, so the scattering occurred because of the repulsion between the two positively charged objects. The known laws of the electric force apply in the interior of an atom — same charges repel each other, opposites attract.

Had Rutherford done his experiments with electron beams, he would have seen a similar scattering phenomenon, but the scattering would have been the result of electrical attraction between the electrons and the nuclei. At that time, however, there was no way to create a beam of electrons with an energy of 5 MeV.

According to Rutherford's atomic model, an atom consists of a shell of electrons that orbit a nucleus. The electric charge of an atom is zero because the positive charge of the nucleus is exactly canceled by the negative charge of the electrons in the shell. The positive charge of the nucleus is due to the positively charged particles, the protons, that it contains.

The simplest atom is the hydrogen atom, the shell of which contains only one electron, circling the nucleus. The nucleus has a positive charge, the electron charge is negative, so the nucleus and the electron are attracted to each other. It is only because the electron is in constant motion that it is not pulled into the nucleus.

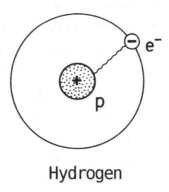

Hydrogen

The hydrogen atom is made up of a nucleus and an electron shell. The nucleus is a proton.

We can picture the hydrogen atom as a microscopic planetary system. In the middle, we have the proton in the place of the Sun, and outside we have an electron representing the Earth.

Newton: It is striking that all atoms have the same structure. Compare any two hydrogen atoms and you will find that they are as alike as peas in a pod. It's strange.

Haller: Yes, this is hard to understand from the laws of classical physics because these laws have no scale. It doesn't matter whether the electron is a millionth of a centimetre from the nucleus, a thousandth, or even a thousand times that distance away. Why does the electron in a hydrogen atom always choose its speed such that it is always the same distance from the nucleus?

Another thing that it is not possible to understand in classical physics is the stability of the atom. One would expect the electron, which is constantly circling the nucleus with its electric charge, to act as a little electromagnetic transmitter, giving off energy in the form of electromagnetic waves. The orbital energy of the electron would diminish accordingly, forcing the electron to fall into the nucleus after only a short time. But there is no sign whatsoever that this happens. There must be some mechanism that keeps the electron stable in its orbit.

Quantum mechanics gives us this mechanism. It provides a theoretical framework that allows us to describe the phenomena of microphysics. The laws of normal mechanics, which we all understand intuitively, do not work in the world of atoms and of subatomic particles. After the first ideas in quantum theory were developed by the German physicist Max Planck in 1900, it took more than 20 years for the consequences to become clear, as they did particularly through the work of Niels Bohr, Arnold Sommerfeld, Werner Heisenberg and Wolfgang Pauli. And it is still a mystery why quantum physics is so successful in describing microphysics.

Richard Feynman, my former colleague at the California Institute of Technology and one of the leading quantum physicists, would often say that nobody understands quantum theory. Since no one understood quantum physics better than Feynman, this was an insult to the rest of the world.

One of the founders of the theory, Niels Bohr, used to say that no one understood quantum theory unless they felt dizzy at the same time. These are both men who contributed much to the theory. They must have felt dizzy all the time!

Einstein: I, too, feel dizzy when I think about quantum theory.

Haller: We simply do not have a better theory. It's true that to understand quantum theory we must give up the intuitions that we have developed in the course of our life. Concepts that we use every day and that are firmly rooted in our perception of reality become suddenly meaningless.

In the realm of microphysics, things that are not considered possible in classical mechanics can happen. We are not capable of comprehending the dynamics of microphysics with the concepts that we have developed in the course of evolution. Quantum mechanics makes it possible to calculate the details of the processes taking place in atomic physics, but it may not be possible to gain a deeper understanding. You are going to have to get used to that, Mr. Einstein. Our world of experience is shaped by classical physics. We are not quantum physicists by nature. We can never really deeply understand quantum physics.

Einstein: Mr. Newton, your brain and mine are too small to understand quantum theory.

Haller: That may well be so, but all our brains are too small, not just yours. The brains of Einstein and Newton are, let's not forget, some of the best in the world.

One of the essential statements in quantum mechanics is that the velocity and the location of an electron can never be measured exactly. There will always be an uncertainty, set by the uncertainty principle that was first found by Werner Heisenberg.

A consequence of this new interpretation of the dynamics inside an atom is that we cannot exactly describe the atomic processes. We can only state the probability that a certain process occurs. It is impossible to know simultaneously and exactly both the location and the velocity of an electron. If we are looking for a precise measurement of the location, we will lose precision on the velocity. The size of the uncertainty is given by Planck's constant h, as Heisenberg discovered.

Einstein: That's terrible. If you are right, God is nothing more than a gambler. I don't believe it. God doesn't play dice.

Haller: This has nothing to do with God; we are talking about atomic physics. The uncertainty relation also applies to macroscopic bodies, such as a moving car, but the uncertainty demanded by quantum mechanics between location and velocity in such situations is so tiny that it can be ignored. And this, ultimately, is the reason why in our intuitive perception of the natural processes the quantum nature of reality is left out.

In atomic physics, however, this is not possible. It is precisely this uncertainty that determines the size of the hydrogen atom. In an atom, the uncertainty in the location of an electron is equal in magnitude to the diameter of the atomic shell, about 10^{-8} cm. Now let's suppose that we are looking at a hydrogen atom whose shell is much smaller than normal, say one hundred times smaller. Now the electron is far more localized than it would be in a normal hydrogen atom. The uncertainty in the velocity of this electron, according to the uncertainty principle, must then increase a hundredfold, making it necessary for the electron's velocity to be much higher in the small atom than in a normal atom.

A higher velocity means a higher energy, so the small atom would have a higher energy than a normal atom. There is an important principle of nature that speaks against this happening. It holds that every system, including atoms, will try to be in the state with the lowest possible energy. In our example the small atom would not be stable. It would not take long for it to expand, radiating energy, until it reached the size of a normal atom.

Newton: Then what happens to artificial atoms that are bigger than normal atoms? Do they become smaller?

Haller: That's exactly what happens. Imagine an atom that is one hundred times bigger than a normal atom. To produce such an atom, we would have to pull the electron away from the nucleus, so it would take energy to manufacture such an atom. Once again, the energy of the new atom is higher than that of a normal atom. This big atom will then shortly return to its normal state, which is the atomic state with the lowest energy. One can't force the electron to give up any more energy once the atom is in this state. The size of the atoms is therefore determined, amongst other things, by the uncertainty relation between the location and the velocity.

So you see, Mr. Einstein, quantum theory manages to explain the size of the atoms. No other theory can do that!

The uncertainty principle, to be exact, does not refer directly to the velocity but rather to the momentum, which is the product of velocity and mass. For this reason, the size of the atomic shell depends on the mass of the electron. If this mass were a hundredth of what we observe in nature, then the atomic shell would be one hundred times bigger, so about a millionth of a centimeter wide. If the mass was a thousand times greater, the diameter of the atomic shell would be only a hundred times bigger than the diameter of the nucleus.

Einstein: Interesting, thus quantum theory brings an essential element of stability to nature. Nature's tendency to produce the same forms over and over again whether in atomic physics, in chemistry or in biology, can only be understood in the framework of quantum physics. That makes quantum physics more likable to me, even if I can't quite imagine a dice-throwing God.

Haller: In this sense, the quantum theory is very important for our understanding of the world. According to classical physics, one could have arbitrarily small atoms, but not in the quantum theory. It is responsible for the phenomena in atomic physics on the typical atomic scale of about 10^{-8} cm. The size of this scale is, however, not really understood. It depends on the mass of the electron, which we can measure but do not understand. If the mass of the electron were a hundred times larger, the atoms would be a hundred times smaller.

Due to the uncertainty principle in quantum mechanics it is impossible to follow the movement of an electron around a nucleus. In quantum mechanics it does not even really make sense to speak of an orbit. All we can say is with what probability an electron will be in a certain part of the space around the nucleus.

This probability distribution of the ground state of an atom does not look at all like an orbit. Rather, the distribution is symmetric about the nucleus, having its maximum in the center of the atom. This probability distribution is given by the wave function of the electron, which can be calculated precisely with the help of the equations of quantum theory. The wave function describes the state of the atom. In quantum theory it is impossible to say anything exact about facts, one can only be exact about probabilities.

Einstein: Another idiosyncrasy of quantum physics that also manifests itself in particle physics is the existence of excited states. If we increase the energy in a hydrogen atom by irradiating it with electromagnetic waves, for instance, the electron enters a new state that corresponds to a higher energy. These states are called excited states and they have very specific energies. We speak of a discrete energy spectrum. The lowest energy state is referred to as the ground state of the system.

Haller: When the atom is excited, it jumps to a higher energy level, stays there for a short while, and then jumps back down. The energy that is released when the atom jumps down is radiated in the form of light or some other type of electromagnetic radiation (X-rays, for example). In order for such an excited state to occur, the atom must receive exactly the amount of energy that is needed for the excitation.

If we want to unequivocally describe the state of the atom, i.e., its wave function, knowing the excitation energy turns out not to be enough. One needs further information, in particular about the angular momentum of the atom.

In the ground state the atom can be turned in any direction and one notices no change because the wave function is symmetric around the nucleus. Excited states can, however, also have angular momentum. The angular momentum can only take certain discrete values, like the energy. But the case of the angular momentum is a little bit simpler than that of the energy. The possible values for the angular momentum are multiples of the smallest angular momentum, which is given by the quantum action h and is often described by the symbol $\hbar$, where $\hbar = \frac{h}{2\pi}$. So the angular momentum can vanish or be $\hbar$, $2\hbar$, $3\hbar$, etc. In atomic physics the symbol $\hbar$ is often omitted and we speak simply of an angular momentum 0, 1, 2, etc.

The ground state of a hydrogen atom has zero angular momentum because the wave function of its single electron is spherically symmetric, so that no spatial direction is any different from any other. This fact alone makes it clear that it will not be possible to understand the hydrogen atom in classical physics. From a purely classical point of view the electron should orbit the proton. The orbital movement of the electron would, however, mean that the electron is rotating around the proton, and so it would have a certain angular momentum. The angular momentum would only be zero if the electron were located right next to the proton and it didn't move. This is not possible according to the uncertainty principle. If the electron were restricted to a very small part of space, its momentum and thereby its velocity and energy would have to be very large. Then the atom would no longer be in its ground state.

In the 1920s physicists came to the conclusion that electrons were more complicated than had previously been assumed. Until then it had been thought that the characteristics of an electron were relatively simple to define. It was considered a point-like object with a certain mass and a well defined charge. When in motion it would have a certain velocity, so it would have momentum and a corresponding energy. At rest, the electron's energy would be that given by its mass, according to Einstein's equivalence between mass and energy, $E = mc^2$. But it was found that electrons have a further characteristic, a kind of intrinsic angular momentum. Here, once again, we have a conflict with classical physics.

If we imagine the electron to be a small sphere, like a tennis ball, it could very well have angular momentum without being in motion. It could be rotating around any internal axis and it would then have, as they say, an intrinsic angular motion. The axis of rotation would describe the direction of the angular momentum, which, just like momentum or velocity, has a magnitude and a direction. Mathematically speaking, these are vectors.

If we now make the radius of the sphere smaller and smaller without changing the angular momentum, the sphere would rotate faster and faster, as happens when an ice skater doing a pirouette pulls in their arms. Imagine reaching the point where the radius of the sphere has shrunk to zero, so the sphere has degenerated into a point, while the spin angular momentum stays unchanged. This is only possible if the rotational velocity becomes infinite.

By considering this limiting case, we can imagine that the intrinsic angular momentum of a point-like particle can take a certain value. We have seen that quantum theory allows only discrete values of angular momentum in an atom. It is the same story for the intrinsic angular momentum, which has been given the special name "spin". The spin can take on only certain values. For the electron, it was found that its spin is not zero but exactly half as large as the value $\hbar$ discussed above. It is $\frac{1}{2}\hbar = \frac{h}{4\pi}$.

One might ask oneself whether the spin of an electron is a kind of rotational momentum that has something to do with the inner structure of the particle, as if the matter inside the electron were turning.

Investigations in this direction have turned out to be nonsensical. In contrast to familiar forms of angular momentum, the spin of an electron is a purely quantum property, corresponding to nothing in classical physics. Like the mass or energy of an electron, the spin is an inner characteristic of the particle.

The intrinsic angular momentum of a classical object (a rotating sphere, for example) can be reduced to nothing by slowing down its rotation. This is not possible with spin. Spin is not a property of electrons that can be switched on and off, it is permanently associated with the particle. An electron always has a spin. It has become a convention in particle physics to leave out $\hbar$ when describing the spin. We simply say spin $\frac{1}{2}$, and we call the electron a spin $\frac{1}{2}$ particle.

Though spin and orbital angular momentum are not the same thing, one thing they have in common is that they both have a direction. The spin of an electron at rest can point in any direction. Here, quantum theory comes into play once again. It states that it is sufficient to consider just two possibilities for the spin, namely that it points either in a particular direction, call it up, or in the opposite direction. In the first case the spin is $+\frac{1}{2}$ and in the second case $-\frac{1}{2}$. We therefore speak of two different spin states. If the spin is neither up nor down, but pointing in some other direction, we can construct the state of the particle from these two spin states.

An electron in the shell of an atom can be characterized by two variables: orbital angular momentum and spin. The first of these takes values such as 0, 1, 2..., the latter is either $+\frac{1}{2}$ or $-\frac{1}{2}$.

The fact that these values are discrete numbers and not arbitrary values is a result of quantum theory. The whole- and half-integer values that we just mentioned are called the quantum numbers of the electron.

The two spins of an electron, pointing up or down.

In the first quarter of the 20th century the physicists found another peculiarity of quantum physics that has no parallel in classical physics. This happened while they were investigating more closely the electron shell of atoms. They did not expect there to be any problem with having an atomic shell with two electrons of the same quantum numbers, let's say an orbital angular momentum of zero and spin $+\frac{1}{2}$.

But there is a problem. There is a law of quantum physics, discovered by Wolfgang Pauli, which states that two electrons in an atom cannot have the same quantum numbers. This rule is called the Pauli exclusion principle. It is a consequence of the fact that electrons have half-integer spin. If the spin were 0 or 1, which is possible, albeit not the case in nature, it would be a different story. Then two electrons could have the same quantum numbers.

The two nucleons, the proton and the neutron, also have spin $\frac{1}{2}$. The Pauli exclusion principle applies to both of them as well. Two protons or two neutrons in the nucleus of an atom cannot have the same quantum numbers. This is the explanation behind some important properties of the atomic nucleus.

Newton: Are there particles that have integer spin?

Haller: Yes, particles with integer spin, e.g., spin 0 or 1, do exist but not as stable particles. They are called mesons. Such particles are unstable objects that are created in collisions and decay soon after they are produced. We don't know of any stable integer-spin particles except the photon, the particle of light.

Let us look again at the atomic nuclei and electrons. Do they have typical inner scales, such as an elementary radius? How small is an electron?

To answer this question, we have to move away from atomic physics. This is a question for particle physics that we still, to this day, cannot answer definitively. But we will return to this later.

Einstein: There's one more thing that should be mentioned: the formulation of the Dirac equation by Paul Dirac in England. This equation was very important for atomic physics, wasn't it?

Haller: Indeed. And that equation was also very important for particle physics. It started with the discovery of the spin. Where does the spin come from? Why is it exactly half as large as the smallest orbital angular momentum that can be measured in atoms?

Dirac, who was an electrical engineer rather than a physicist, introduced a new mathematical quantity called a spinor, which is something like a vector. A spinor is like a vector in that it changes when turned around in space, but a spinor changes in a different way. A lot of work has been done on this mathematically. Dirac, who was not a mathematician, arrived at the concept of a spinor in a very intuitive way. His spinors consisted of four complex numbers.

Dirac also noticed that his equation predicted both negative and positive energies. Dirac suggested that for every particle there should be an antiparticle. The antiparticles are particles with negative energies.

A short while later, the first antiparticle was discovered here at Caltech by Carl Anderson. It should be stressed, however, that Anderson did not make his discovery because of Dirac's prediction. He did not even know about it. He found the positron by examining the tracks of particles in a cloud chamber. He found a track that looked like that of an electron, but made by a particle that had a positive electric charge. That was the positron, the antiparticle of the electron.

This experiment made it clear that the Dirac equation was a goldmine. Today we know that many aspects of particle physics follow directly from this equation, since the particles of nature such as electrons and quarks can be described by the Dirac equation.

But back to the electron. The electron is the most important elementary particle. An electric current is produced by electrons. Yet, ever since its discovery, the electron has presented us with riddles. Because it has a mass, the electron was thought to have a finite size, and therefore a finite radius. Experiments are carried out with ever increasing precision but they still give a result of zero. It seems that the electron is an elementary object *par excellence*, a mathematical point that is endowed with the attributes of charge, spin and mass. Should an electron have an inner structure, we now know, at least, that it has to be smaller than 10^{-17} cm. That is a billion times smaller than the diameter of an atom.

Newton: Is that even possible? Can there be such a thing as a point-like object with mass and charge and spin? In my opinion, that is a bit too much for a point. A point with charge, spin and mass is a pretty strange thing.

Einstein: I agree. The mass of an electron is a physical quantity. It must come from somewhere. A mathematical point cannot have a mass.

Haller: Yes, I also have my doubts. From a mathematical point of view, a point is what is left when you take a small sphere and let its radius go to zero, so it is the result at a limit. If you imagine a small sphere having the mass and charge of an electron and then let its radius go to zero bit by bit, you have the electron. It is a little like Lewis Carroll's famous Cheshire cat, which slowly disappears until only its smile is left, a smile without a cat. We will have to wait and see whether future experiments with electrons will change this point of view. The issue is still open.

Before we attack the atomic nucleus, I'd like to suggest another short coffee break.

Einstein: Good idea. Let's go up to the restaurant and have something to drink, and let's not talk at all anymore about quantum theory.

So the three physicists adjourned to the restaurant. Haller went to the physics department quickly to discuss something with the secretary. Then he came back to the restaurant to enjoy the excellent Italian coffee with Einstein and Newton.

ATOMIC NUCLEI AND PARTICLES

4

After the break the three physicists met again in the library of the Athenaeum.

Haller: Now we start to discuss the atomic nuclei. The nucleus provides the greatest part of the mass of an atom and has been proven to be, like the atom, a system of smaller interacting particles. We now know that the atomic nucleus is composed of two types of particles, positively charged protons and electrically neutral neutrons. The mass of the proton is 938.272 MeV. That makes the proton 1836 times heavier than an electron.

Newton: So more than 99.9% of the mass of the hydrogen atom is concentrated in the nucleus? Is this number 1836 also a natural constant?

Haller: Yes, nobody has been able to calculate this number. Nowadays we refer less often to the proton mass than to an artificial mass, which we obtain by dividing the mass of a carbon nucleus by the number of nucleons that it contains. The mass that we get, 1.660540×10^{-27} kg, is almost the same as the proton mass. I have given the mass in kilograms here even though that is not the custom in particle physics. The mass of an electron is 9.109390×10^{-31} kg. Why the ratio of the electron mass to the proton mass is so small is unknown.

The electric charge of the proton is the same as the electric charge of the electron, only with the sign reversed. Therefore the hydrogen atom has no charge. We do not understand this phenomenon either. One could easily imagine the proton having a charge other than the charge of the electron multiplied by -1, but then the hydrogen atom would have an electric charge.

The fact that the charge of the hydrogen atom vanishes is well known. We can give a very precise limit for the possible charge of the hydrogen atom. Compared to the charge of the proton, e, the hydrogen charge must be 10^{-21} times smaller.

It is remarkable that the electric charge of the proton and the electron are exactly the same, even though the physical properties of these particles, e.g., their masses, are so different.

Newton: Obviously there is a connection between the electron and the proton which forces their charges to be the same except for the sign. It is curious that this should be the case, since the proton is made up of quarks but the electron is not. The connection should actually be between quarks and electrons, I think.

Haller: Yes, that is what we believe today, although it is not at all clear what is really going on here. Later we will see that a connection between quarks and electrons might exist. But this will bring us far beyond atomic physics to the "Grand Unification" of all interactions.

By the way, an exact measurement of the ratio of the proton mass to the electron mass gives the number 1838.683662.

Newton: It is crazy that we can measure this number so precisely and still no one has any idea where this number comes from.

Haller: Yes, that is frustrating. But there is something else of interest concerning the mass of the electron. I had to give a lecture here at Caltech once, years ago, in 1989, at a conference organized on the occasion of the 60th birthday of Gell-Mann.

I asked Gell-Mann what I should speak about. He laughed and gave me a free hand in choosing the topic, but he suggested that it might be good to say something about the mass of the electron. That was, of course, a joke, because he knew just as well as I did that no physicist could say anything intelligent about the mass of the electron. He might as well have asked me to say something about astrology.

However, while I was preparing my lecture, I took a closer look at the mass. I also asked a colleague who was in the office with me, an experimental physicist from Zurich, Valentine Telegdi, what I should say. He told me to simply state the mass of the electron in different units and leave it at that, because that was all one could say about the electron mass. Thus I calculated the electron mass in various units, including the American pound.

When I calculated the electron mass in pounds, which probably no physicist before me had ever done, I got a surprising result. My calculator showed a whole number. It turns out that, disregarding powers of ten, the electron mass in pounds is to a very good approximation a whole number.

$$m_e = 2 \times 10^{-30} \text{ lb}.$$

Einstein: That is interesting. I always found the pound, which is 453 grams, a little strange. Now I like the pound. After all, the electron mass would be very simple if we were to use pounds as our unit.

Haller: All atoms except the hydrogen atom possess complex nuclei that contain more than one particle. This starts with a special kind of hydrogen called heavy hydrogen, whose nucleus is twice as heavy as that of normal hydrogen but has the same electric charge. This is because its nucleus contains besides a proton also a neutron.

The English physicist James Chadwick discovered the neutron in 1932. The mass of the neutron is 939.565 MeV. Thus the neutron is only a little more than a tenth of a percent heavier than the proton. Not only do protons and neutrons have almost the same mass, but they are also related by a force which is found only in the atomic nucleus. This force, many times stronger than the electric force between the electrons and the nucleus, is called the strong nuclear force.

Werner Heisenberg was interested in the relationship between the two particles. He introduced a symmetry that links the proton and the neutron and which is known in nuclear physics as the isospin symmetry. This symmetry is no accident. As we will see, it has to do with the special dynamical properties of the quarks, in particular with the masses of the quarks. But more about that later.

The fact that the neutron is a little bit heavier than the proton must be counted among the facts of microphysics that have not been understood in detail. Intuitively, one would expect the opposite. The proton is electrically charged, so it is surrounded by an electric field. This field has an energy which, according to Einstein's energy-mass relation, must contribute to the mass of the proton. Thus one would expect that the proton is heavier than the neutron. We will return to this problem soon because it is connected to the inner structure of the nuclear particles.

Neutrons and protons, also often called nucleons, are the building blocks of all atomic nuclei. The number of nucleons in a nucleus depends on the element at which we are looking. The atomic nucleus of carbon, for example, normally has six protons and six neutrons. Heavy atomic nuclei always have more neutrons than protons. So we find in the nucleus of uranium, for example, 92 protons always and most often 146 neutrons.

The repulsive electromagnetic force in the nucleus could easily make the nucleus explode if it were not for the other force, the strong interaction, which holds the neutrons and protons very close together. The particles are squeezed together by the strong interaction.

Atomic nuclei are of the order of 10^{-13} cm in size. A nucleus is about 100,000 times smaller than the atom. If we were to imagine the nucleus of an atom to be as big as an apple, then the whole atom would have a diameter of a considerable size, say around 10 km.

The nucleons have a finite radius that is of the same order as the size of the nucleus, around 10^{-13} cm. We measure the size of the nucleons by scattering electrons off them. When an electron flies past a nucleon, its flight path is deflected according to the distribution of charge inside the particle. That is how we know that the charge of a proton is not concentrated in a point, but distributed over a sphere of radius 10^{-13} cm. We have also discovered that a neutron, although it has no electric charge, has charges inside which cancel each other.

At the beginning of the '50s, many new particles were discovered in cosmic ray experiments and with accelerators. All of the particles were unstable. By the end of that decade particle physics was like a big zoo, filled with dozens of new particles. It was a big mess. But eventually it was possible to bring order to this chaos. This started at the beginning of the '60s. Murray Gell-Mann in the United States, and his Israeli colleague Yuval Ne'eman, who wasn't even a physicist (he worked at the Israeli embassy in London), discussed a new kind of symmetry based on the mathematical group SU(3). Soon it was seen that this symmetry was capable of describing the particles.

The most important conceptual breakthroughs in fundamental physics in the second half of the last century were gained thanks to experimental studies using the big accelerators as well as through theoretical research. These breakthroughs concerned the clarification of the structure of matter and the laws of the microcosmos. It was discovered that the nuclear particles and the newly discovered particles could be described in a simple way using three basic building blocks, the quarks.

This discovery was made at Caltech. Murray Gell-Mann developed the quark model here in 1964, while George Zweig worked on it at CERN in Geneva. Zweig was a younger colleague and ex-student of Gell-Mann. Gell-Mann was able to show that the quark model easily produced the same symmetry results that he had discussed three years earlier, in 1961, on the basis of the symmetry group SU(3).

It was also Gell-Mann who coined the word "quarks". He took it from *Finnegans Wake* by James Joyce. The story goes that James Joyce heard this word while passing through Freiburg in Germany. In the town's market square tradeswomen advertised their curd (*Quark* in German), and it is said that Joyce found this word so amusing that he used it in his book.

Einstein: Yes, I remember the book. It refers to three quarks for Muster Mark. So three quarks are mentioned, and Muster Mark would be Gell-Mann himself, I guess.

Haller: That could be, even if Gell-Mann would never admit it today. That quote, by the way, is from page 383 of the English edition of *Finnegans Wake*, which is remarkable because both the numbers three and eight play a special role for quarks. The name that Zweig used — aces — found no acceptance.

One needs two different quarks, referred to by the symbols u for "up" and d for "down", for protons and neutrons. A proton is composed of two u quarks and one d quark: p = (uud). In the case of the neutron, the roles of u and d are interchanged: n = (ddu).

The electric charges of the quarks are particularly interesting. Between u and d there must be a difference in charge of one. Thus we can quickly calculate what the charges must be.

The charge of the u quark is two thirds that of the proton, that is $+\frac{2}{3}$, and the charge of the d quark is one third that of an electron, namely $-\frac{1}{3}$. The charge of each of the two nucleons is the sum of the quark charges. For the proton we get: $\frac{2}{3}+\frac{2}{3}-\frac{1}{3}=1$. The charge of the neutron is: $\frac{2}{3}-\frac{1}{3}-\frac{1}{3}=0$. So there are indeed electrically charged objects, the quarks, inside a neutron, and their charges exactly balance.

Without a doubt, these strange charges are a very surprising property of the quarks. The quark hypothesis was severely criticized for this reason. Very few physicists were willing to accept these charges. Zweig even had problems getting his work published. It still hasn't really been published because Zweig refused to publish his paper in a European journal, and he was not allowed to send his paper to an American journal.

Gell-Mann did not want to send his paper to the American journal *Physical Review Letters*. Instead he sent his paper to *Physics Letters* in Europe. The journal's editor at CERN had doubts about publishing the paper because of the strange charges. But he did finally publish it, making the argument that it was Gell-Mann who was risking his good reputation in professional circles and that since it was his own risk, not the journal's, they may as well publish his nonsense.

And that is how Gell-Mann's paper was published in *Physics Letters*. It became probably the most important paper ever published in that journal. In my opinion those (almost) two pages that Gell-Mann wrote for *Physics Letters* contain more physics per page than any paper in the physics literature since the time of Isaac Newton.

Newton: Great, I am going to read this paper soon. The charges, I must say, are truly extraordinary. Is there a way to better understand them?

Haller: We don't really understand the charges, although we will soon find out that there is a new symmetry for quarks which at least makes the peculiar charges plausible.

Nucleons and the unstable mesons, the pions, are not the only particles that we can build from quarks. The delta particles, often called the delta resonances, also consist of quarks. The delta particle with electric charge

+2 has the substructure (uuu). Its spin, $\frac{3}{2}$, is easily calculated by adding up the spins of its three quarks, which in this case are all pointing in the same direction. Exchanging u for d, one notices that there are four delta resonances, the others having the following substructures: (uud), (udd), (ddd). Their electrical charges are, respectively, $(+2, +1, 0, -1)$. These particles form an isospin family, composed of four members.

The delta particles are excitations of the proton which can be produced by bombarding nucleons with π mesons. This is how they were discovered in the experiments at the accelerator in Berkeley. They decay very quickly — in about the length of time that it takes a particle of light to cross an atom's nucleus — into mesons and nucleons. For something with such a short lifetime it is questionable whether the term particle is really fitting. A particle's main characteristic is its mass, and the mass in this case cannot be given exactly.

Quantum theory tells us that there is an uncertainty relation between the mass of a particle and its lifetime. The shorter the lifetime of the particle, the greater the uncertainty in its mass. In the case of the delta resonance we cannot give the mass precisely. The best one can do is to specify an average mass, which one obtains by averaging over many observed delta particle decays.

The averaged mass of a delta particle is 1232 MeV. The uncertainty in this mass, known as the decay width, is 120 MeV, so about 10% of the mass. They decay very quickly. It is for this reason that we often call the delta particles the delta resonances, because each one is really more a resonance than a particle.

Both the (uud) delta resonance and the proton have the same quark structure, so one must ask oneself what the difference is between these two particles. It is simply a matter of the spins. In the delta resonance, the spins of the u quarks and the d quark point in the same direction, this is not the case in the proton. To turn a proton into a delta resonance, one would have to flip the spin of one of the quarks. This however takes energy; the amount of energy corresponds exactly to the difference in their masses, almost 300 MeV.

Especially because of the unfamiliar electrical charges, some physicists originally assumed that quarks were not really elementary building blocks for particles but abstract mathematical entities used to describe the particles and their symmetry properties. Gell-Mann himself was of this opinion. However, with the help of the electron accelerator at the Stanford Linear Accelerator Center (SLAC) at Stanford University, it was discovered that indeed there were hard, seemingly point-like charged objects inside the proton.

At SLAC, physicists sent electrons that had been accelerated to almost the speed of light to collide with atomic nuclei. The electrons' flight paths were usually deflected only slightly. But to the surprise of the physicists, once in a while an electron would drastically change its direction, just as the alpha particles had done in the Rutherford experiment. Thus the

View of SLAC, Stanford University.

electrons, on their way through nuclear matter, must sometimes collide head-on with point-like charged objects.

By examining the data more carefully, it was possible to draw conclusions about the electrical charges inside the proton. They turned out to be $+\frac{2}{3}$ and $-\frac{1}{3}$, so the particles looked like quarks.

I was working with Gell-Mann at the time, and I mentioned the possible connection between the SLAC experiments and quarks to him a number of times, but he didn't take it seriously. For a long time, Gell-Mann showed no interest in the results from SLAC. But then I showed him how one could find these results without the quarks being actual particles, and suddenly he developed an interest in my approach as well as in the experiments at SLAC. We began working on these problems together.

Although the experimental data gave us no option except to believe that the quarks were receiving hard blows in collisions with the electrons, there were no free quarks found in the debris. Apparently the forces between the quarks were so strong that it was not possible to separate them.

A more detailed examination of the data produced a further riddle for the physicists. There are two parameters that are important when analyzing a collision between an electron with a large momentum and a proton. One is the scattering angle of the electron which could, for example, be 20°. The other is the energy lost by the electron in the collision. The energy of the electron could, for example, be 20 GeV initially and after the collision be only 12 GeV. The energy and angle distributions can be exactly measured. They tell us something about the distribution of the momenta of the quarks in a fast-moving nucleon.

Einstein: When one pictures a proton as three quarks, one might assume that a fast-moving proton with an energy of, say, 18 GeV behaves like a bundle of three quarks, each with an energy of 6 GeV, a third of the total energy. Accordingly, the momentum of one quark would be one third of the proton's momentum.

Haller: Yes, one might expect that, but the experiments show a different picture. The quarks do not carry a third of the total momentum, they show an interesting momentum distribution. Sometimes a quark does have a third of the total momentum but often its momentum is a lot less than this, one tenth of the total momentum, for example. This information is of great interest in terms of our theoretical understanding because it tells us something about the forces between the quarks.

To the astonishment of the physicists, a quark's average share of the total momentum was much lower than expected. After measuring many details, it turned out that the sum of the momenta of the quarks was not the total momentum, as expected, but much less, around 50% of the total. So the quarks contribute only about half of the momentum of a fast-moving nucleon. The question is: where does the other half come from?

Einstein: That means there have to be further constituents. Maybe these are the particles responsible for the forces between quarks?

Haller: Mr. Einstein, you are right. I offer you my compliments. There must, in addition to quarks, be other constituents that are not observable in a nucleon–electron scattering experiment. Since these experiments only register electrically charged constituents, we conclude that the other building blocks must be electrically neutral. So they contribute only to the momentum, not to the charge.

It will become clear that these neutral particles or quanta really do have something to do with the forces between quarks. The name of these particles describes their role. They are called gluons, and they are the glue that holds three quarks together to form a nucleon.

If we disregard the gluons, the nucleons consist of just two different types of quarks: u and d. In the scientific literature, the term "flavor" is used to describe these labels. We speak of the "quark flavors" u and d.

These two quarks are the elementary building blocks of the atomic nucleus, but the term "building block" must be interpreted in a special way.

Normally we would speak of building blocks if we expected to be able to break the nucleon apart into its constituents. But to this day no one has succeeded in breaking a nucleon into quarks, and most likely no one will succeed in the future. In dealing with the substructure of nucleons we seem to have reached a level of substructure of matter where our notions, which are based on our experiences in the everyday world, become invalid.

If we include the electron as the building block of the atomic shell, we can say that the normal matter in the universe consists of electrons, u quarks and d quarks. But beyond the u quarks, d quarks and electrons lies a new world of unstable particles, a world that can only be explored with the help of accelerators.

The first step into this new world was made in 1937, not with an accelerator but with the help of a particle-detecting apparatus called a cloud chamber. It allows us to see traces of electrically charged particles in cosmic rays.

The upper layer of the Earth's atmosphere is constantly being bombarded by fast-moving particles coming from the far reaches of the universe. These particles are mostly protons or light atomic nuclei such as deuterons or alpha particles. The collisions that occur with atomic nuclei in the air can produce short-lived particles that have an electric charge equal to that of an electron and which decay in about two millionths of a second.

These new particles, called μ particles, have a mass of about 105.7 MeV, making them about 200 times heavier than an electron. Further experiments, which were not carried out until the '40s, showed that we are dealing with a particle that has charge but no structure. This particle is the heavy brother of the electron. The only difference between this particle and an electron seemed to be the greater mass and the fact that these new particles were unstable.

The μ particles, today often referred to as muons, apparently do not play any role in the structure of normal matter. The muon seemed to be a totally useless particle. The American physicist Isidor Rabi once asked: "the muon, who ordered that?" As it turned out, nobody had ordered it. But there it was.

To this day, no one has an answer to Rabi's question. But apparently muons do play a role in the physics of particles, and we will examine this role in more detail later on.

Einstein: I seem to remember that there was a problem with muons. These particles have a short life span of only about 2.2×10^{-6} s. It was hard to understand how they could get to the Earth's surface at all. They should have decayed before they reached it. My theory of relativity helped to explain this.

Haller: You are completely right. In your theory we have the property of time dilation, which has the consequence that a fast-moving muon can fly longer than we would naively calculate. Careful observations showed a complete consistency between your theory and the observations. Your theory of relativity, Mr. Einstein, turned out to be right.

I wanted to mention something else that is closely connected with your theory. Normally we do not encounter positrons (the antiparticle of the electron) in nature but it is easy to produce them in particle collisions, for example in the collision between two photons. Here an electron–positron pair is created. This is probably the simplest reaction in which the energy-mass equivalence that you, Mr. Einstein, discovered in 1906, can be seen. From the energy of the two photons, two particles with mass are created.

The destruction of the positron is also impressive. It grabs an electron from an atom in its neighborhood, and the electron and positron are turned into two photons, sometimes three. Matter turns itself into light.

Einstein: Indeed that is a very nice example of my energy-mass relation, probably the nicest one of all. I could not have thought of it in 1906 since I did not know that antiparticles exist. I thought more about radioactive processes, in which a small part of a mass is turned into energy.

Haller: Looking at my watch, I see that it's already almost noon. The next thing that we have to talk about are the forces between the quarks. I suggest we do that in the afternoon. Now it's time to think about lunch. We should go to Chandler Hall, that's where most people from Caltech go.

The three physicists left the Athenaeum and headed west. After about 200 meters they reached Chandler Hall, where students and scientists from Caltech often ate. Today was pizza day, and the three physicists sat down at a table to eat ham and pepperoni pizza.

Haller then went to get dessert, which was a serving of lemon ice cream and an espresso. Newton quite obviously enjoyed the espresso, as he hated American coffee. The latter was, he said, nothing more than slightly brown-colored water. Einstein and Newton did not dissent, but they didn't mind American coffee. They could drink it like water, as much as they wanted, and still they wouldn't get a stomach ache. That was not possible with strong French or Italian coffee.

BIG ACCELERATORS

5

In the early afternoon the three physicists met once again in the library. Einstein started the discussion with a request.

Einstein: Mr. Haller, both Newton and I are theoreticians, and we have little knowledge of how new insights have been obtained, especially in particle physics. I suppose that the experiments were done with accelerators. So I suggest that in our session today, we talk about accelerators first, then discuss the theoretical issues afterwards.

Haller: That's fine. I am a theoretical physicist, too, of course, but my work is relatively close to the experiments. Some experiments were even suggested by me. Experimentalists sometimes invite me to participate in their projects, and my name is sometimes mentioned in their publications. Thus let's talk first about accelerators.

Rutherford used alpha particles to examine the structure of atoms. Alpha particles were easy to get because they are emitted by radioactive nuclei. However, if one wants to examine the inner structure of the nucleus and of the nucleons, alpha particles are of no use. First, alpha particles are nuclei themselves, nuclei of helium, and second, alpha particles typically have an energy in the region of 5 MeV, which is much too low. It may sound like a paradox, but it is a simple consequence of quantum theory that the smaller the structure being examined, the larger the energy or momentum of the particles used as probes must be, and that, in turn, means a larger apparatus. According to the uncertainty relation, a higher momentum means a smaller uncertainty in location and vice versa.

It is easy to calculate that particles with an energy of a couple of MeV, such as the alpha particles in the Rutherford experiment, can reveal structures of the order of 10^{-12} cm. But that is the limit of their resolution. Higher energies are needed if we are to examine smaller structures. Particle physics therefore needs accelerators in which stable particles such as electrons or protons can be accelerated to high energies. In fact, particle physics is often referred to as high energy physics because of the high energies required.

Modern particle accelerators are complicated and expensive machines, but the basic principles of accelerator technology are simple and easily understood. You could build a simple particle accelerator yourself. You would need a glass pipe a few centimeters long, two metal plates, and a 12-volt car battery. The metal plates have to be mounted at the two ends of the glass pipe and then connected to the poles of the battery. Then you take the air out of the pipe with a pump. Any electron in the pipe that happens to be near the negatively charged plate will be repelled. Accelerated by the electrical force, the electron will make its way towards the positively charged plate and finally collide with it. The energy of the electron on arrival is given by the voltage of the battery: 12 electron volts.

To turn this apparatus into an accelerator, we just have to replace the positively charged metal plate with a screen of thin wires. This still attracts the electrons, but lets most of them pass through. So we now have an electron beam with an energy of 12 eV. The energy of this beam is relatively low, but the velocity of the electrons is considerable.

This apparatus can produce a strong current of electrons if we arrange for a lot of electrons to be freed near the negatively charged plate. This could be done, for example, by bombarding the plate with high energy electromagnetic radiation or by heating up a metal wire in the vicinity of the plate, thereby causing electrons to be ejected from the metal surface. The tube found in some types of television works on a similar principle, but the energy of the electrons hitting the screen is much greater, about 20,000 eV.

Newton: We can calculate the electron's velocity easily since the energy is much less than the mass-energy of the electron, i.e., 511,000 eV. The kinetic energy of the electron, 12 eV, is equal to $\frac{1}{2}mv^2$. Since the mass of the electron corresponds to the energy mc^2, the ratio 12 eV/511,000 eV must be equal to $\frac{1}{2}v^2/c^2$. That gives a velocity for the electron of about 2060 km/s, assuming that the velocity of the electron at the negatively charged plate was zero.

Haller: Yes, that is a simple way to calculate the velocity. In the case of a television, where the energy is 20,000 eV, an electron's velocity is calculated to be 84,000 km/s, more than a quarter of the speed of light. Slowly things are going to get problematic for your theory, Mr. Newton. We're going to need Einstein's relativity theory.

Every accelerator works according to the principle I've just described. The particles are always accelerated by an electromagnetic field. This only works, of course, if the particles have an electric charge. Neutrons cannot be accelerated in this manner. Since it takes time to accelerate a particle, it is also necessary for the particles to be stable or at least to have a relatively long lifetime. Electrons and protons pose no problem, but just about every other particle considered in particle physics decays quickly and cannot be accelerated.

Since atoms are made up of electrons, protons, and neutrons, getting electrons and protons is no problem. It is also possible to accelerate positrons and antiprotons, but obtaining them is quite complicated, as they must be produced in particle collisions. Other particles that can be accelerated by this method include light nuclei such as deuterons or alpha particles, and also heavy atomic nuclei such as the nuclei of lead or uranium.

The 12 eV accelerator that I described only accelerates electrons to a little bit more than 2000 km/s. If we put three further accelerator units behind it, we would reach four times this energy, or about 48 eV, and thereby produce electrons with twice the velocity, over 4000 km/s.

By placing many accelerator units behind each other, we can increase the energy more and more, but the velocity does not increase in the same way as before. At around 100,000 km/s the classical laws of mechanics fail and have to be replaced by Einstein's relativistic dynamics. The closer the particles' speed gets to the speed of light, the harder it becomes to increase their velocity. The accelerated particles reach 99% of the speed of light relatively quickly. But if we increase the energy further, there is hardly any increase in velocity. For example, once the velocity reaches 99% of c, it is necessary to increase the energy by a factor of three to get to 99.9%.

Einstein: That engineers have to concern themselves with my theory of relativity is something that I like to hear. In 1905 I would not have dreamed of it.

Haller: Yes, your theory is old hat for engineers who build accelerators; even some politicians have to concern themselves with your theory. There was a fight many years ago in the US Congress about the budget for high energy physics. It took a long time before most members of Congress agreed to the budget. They had problems understanding why it was necessary to build an accelerator that could accelerate protons to 99.95% instead of 99% of the speed of light, making it substantially more expensive — and for a difference of less than one percent.

But back to accelerators. The Bevatron accelerator, where the antiproton was discovered, has operated in Berkeley, California since 1955. It accelerates protons to 99% of c, corresponding to an energy of 7 GeV. The apparatus was built specifically to discover the antiproton and it worked.

The accelerator AGS in Brookhaven, Long Island, New York, started in 1960 and reached 99.95% of c, corresponding to an energy of 30 GeV. It was very important for the development of particle physics.

The Large Hadron Collider or LHC, the new accelerator at the European particle physics center CERN, will reach an energy of 14 TeV, or 14,000 GeV. That's 15,000 times the mass of a proton in units of energy. The velocity of the proton in this case is $0.999999998c$, so almost equal to c.

Einstein: That makes it clear that the velocity of particles is no longer meaningful in high energy physics. What's important is the energy and the associated momentum, in line with my theory.

Haller: The statement from Newtonian mechanics that momentum equals mass times velocity does not hold in high energy physics. In Einstein's dynamics, energy and momentum are essentially proportional. The constant of proportionality is the speed, which at high energy is approximately the speed of light. Only in cases where the velocity is small compared to the speed of light and when the energy is given primarily by the mass (by $E = mc^2$) does Newtonian mechanics apply. So a proton with an energy of 100 GeV has a momentum of about 100 GeV/c. The momentum of a high energy particle is therefore essentially directed energy.

In 1960, on the premises of Stanford University, the SLAC electron accelerator began its operation. This was a highlight of accelerator technology. SLAC works on the aforementioned principle: electrons are successively accelerated to higher and higher energy over a distance of two miles.

Electrons at SLAC were accelerated at the start to an energy of 20 GeV, and later they were accelerated to an energy of 50 GeV. This gave the physicists at SLAC a gigantic microscope, capable of looking deep into the atomic nucleus. We will deal later with the structures that were discovered in this way.

Einstein: I remember that around 1930 people came up with a way to reach high energies with only one accelerator unit. The trick was to accelerate the particle with the same unit many times.

Haller: Yes, that was an important discovery made at Berkeley. The particles can be accelerated along a curved path, completing a circle so that they return to their starting point and are accelerated once more. This is accomplished through magnetic fields.

In a constant magnetic field, an electrically charged particle will travel in a circular path. After each full circle the particles receive another push, so their energy is constantly increasing. If nothing were done, the particles would leave the circular path after their energy increased. To keep the higher-energy particles on the same path, the magnetic field has to be increased. Theoretically, we could keep increasing the energy of the particles indefinitely, but the strength of the magnetic field is a limiting factor. The magnetic field cannot be increased indefinitely, it reaches a maximum that depends on factors such as the materials and technology used to build the magnets.

A ring accelerator, in which particles pass through a ring-like pipe in a vacuum, is described by two parameters, the radius of the ring and the maximum strength of the magnetic field. The bigger the radius and the larger the maximum strength of the field, the higher the particle energy that can be reached.

In the second half of the 20th century, several accelerators were built in Europe, the United States, Japan, and the former Soviet Union.

Especially remarkable are the Bevatron in Berkeley, California (1954, energy: 6 GeV), the AGS in Brookhaven, New York (1960, 30 GeV), the PS at CERN near Geneva, Switzerland (1959, 25 GeV), the Fermilab accelerator near Chicago (1972, 200 GeV, later 400 GeV), and the SPS at CERN (1976, 400 GeV, with a ring of circumference 26.7 km).

In most of these accelerators the particles were not accelerated in only one ring. It was found to be more efficient to use several rings. At CERN, for example, particles for the SPS were first accelerated in the PS ring, then injected into the SPS ring where they received a final acceleration.

The Fermi National Laboratory (Fermilab) near Chicago.

When it became possible to construct superconducting magnets, the accelerator technology made a considerable leap. In a superconducting material the electric current encounters no resistance. No energy is lost as heat. It is also possible to send larger currents through the wires of the coils, thus creating magnetic fields that reach about 10 Tesla. The currents involved are several thousand amperes. For comparison, the strength of the Earth's magnetic field is about 1/20,000 Tesla, so the strongest magnetic fields are 200,000 times stronger than the Earth's magnetic field. In 1983, with the help of superconducting magnets, Fermilab in the United States managed to raise the energy of the colliding protons to 900 GeV.

A collision between two accelerated protons or between an accelerated proton and an atomic nucleus at rest takes a very short time, roughly the amount of time that it would take the proton to cross the atomic nucleus at the speed of light, or about 10^{-24} s. This makes it impossible to follow the collision itself, but we can study the products of the collision. We know which particles were involved in the collision, and we can see which particles leave the scene and with what energy and momentum.

It's like a traffic accident without witnesses. The police will attempt to reconstruct what happened by examining the facts at hand: the damage to the vehicles, the location of the vehicles, etc. The policemen can rely on the fact that the rules of classical mechanics applied to the accident. The complication in particle physics is that the natural laws that govern particle collisions are not known if one is exploring a new energy region. One collision does not tell us much. We have to examine many collisions in order to learn details about the natural laws of microphysics.

Einstein: But you will only learn things if the theoreticians have concrete theories. Experimental physics is not enough on its own.

Haller: Well, of course, and experimental physicists know this. Although there are some who believe they could do completely without theory, the good experimental physicists pay attention to theory.

The instruments that measure the properties of the particles produced in these collisions are constantly being made more efficient. In the early years of particle physics experimentalists used cloud chambers and scintillation counters. Later they used bubble chambers, in which particles leave tracks in a special liquid. Another type of detector reveals particle tracks with the help of charged metal plates which produce sparks when a particle passes between them. Such detectors are called spark chambers.

Today we almost exclusively use complicated electronic instruments. Their advantage is that the signals they register when a particle passes can be immediately stored by a computer for further processing. That makes the analysis of particle collisions much easier. A little of the romance is gone from particle physics though, because the collisions can no longer be seen with the naked eye.

In particle collisions, as in our everyday world, the laws of conservation of momentum apply. Imagine that a proton with an energy of 1000 GeV hits a proton at rest. The moving proton has a momentum of 1000 GeV/c, the proton at rest has no momentum. After the collision the net momentum must still be 1000 GeV/c. This means that the sum of all the momenta of the departing particles must be 1000 GeV/c. What happens in practice is that all the particles fly forward, in the direction that the incoming proton had been traveling.

The main goal of colliding particles, though, is to turn as much of the available energy into mass as possible. It would be best if it were possible in principle to turn all of the available energy into mass. In the example just mentioned, that would be a mass of 1000 GeV, expressed in units of energy. The conservation of momentum, however, forbids this from happening. Only a small part of the available energy can be turned into mass, up to 42 GeV in this example.

Newton: One could get around this problem by letting the accelerated proton collide with another moving proton rather than with one that is at rest. Thus the total momentum could be zero.

Haller: Yes, the particle physicists started doing that some time ago. We shoot a proton or antiproton at another proton or antiproton that has the same energy but the opposite momentum, such that the sum of the momenta of the two particles is zero. In this case the force of the collision is much stronger. If the energy of each particle is 1000 GeV, there would be 2000 GeV available for the creation of mass.

The analogy with a traffic accident is helpful here as well. The damage to a car is considerably less in an accident if it runs into a stationary car compared to crashing head-on into a car that is moving towards it. But in an accelerator head-on collisions are harder to arrange than collisions with a "target" at rest. You need two particle beams, and it is not so easy to guide a particle beam so perfectly that it will collide head-on with another beam. Look at it this way: to hit a tree trunk with a rifle bullet is relatively easy, but hitting another bullet in flight is much more difficult! In particle physics you care not only about the energy of the collision but also about the number of collisions happening per second. This is called the luminosity.

In spite of the many technical problems, it has been possible to arrange for head-on particle collisions. A machine that can do this is called a "collider". The first collider, the Intersecting Storage Ring (ISR), was finished at CERN in 1972. At the ISR two parallel pipes were filled with protons with an energy of 30 GeV. Thus an energy of 60 GeV was available for the creation of new particles. Collisions occurred at specific collision points. A collision rate of 10 million collisions per second was finally reached after many starting difficulties.

At the beginning of the '70s the physicists at SLAC built a comparatively cheap but very successful collider called SPEAR. Electrons and positrons were sent around in a ring in opposite directions. Since positrons are positively charged, they are deflected in a magnetic field in the opposite direction to electrons. If a positron drifts to the left in a magnetic field, an electron will drift to the right. This makes it is possible to have electrons and positrons flying together in one collider ring, in the same vacuum tube, provided they have the same energies. When an electron and a positron collide, there is a complete annihilation. All the available energy can go into creating new particles. This is why electron–positron collisions are particularly effective when attempting to discover new particles and phenomena. Discoveries made at SPEAR and later at DORIS at DESY in Hamburg played an important role in the development of the Standard Model of particle physics.

The weekend of 9 November 1974 will never be forgotten. The physicists at SLAC observed that the colliding electrons and positrons produced a new heavy particle with a mass of about 3.1 GeV. This particle was later given the double name J/Ψ, and it played an important role in the further development of particle physics. We will come back to this. On that weekend I was at Caltech and working in my office on a Sunday because it was raining. Suddenly my colleague Richard Feynman showed up and told me the interesting news from SLAC. In the following week Feynman reported the new results daily.

The DESY laboratory in Hamburg, Germany.

The double name J/Ψ was chosen because the new particle was discovered simultaneously in Brookhaven, by the group of Sam Ting, and at SLAC, by the group of Burt Richter. They could not agree on a single name for the new particle.

After the success of SPEAR, three other electron–positron machines were built — PETRA at DESY, PEP at Stanford, and TRISTAN in Tsukuba, Japan.

In the early '90s the electron–positron collider LEP at CERN started to operate. This was the first accelerator able to undertake experiments of great precision at high energies. The circumference of the LEP ring was 26.7 km. The research program at LEP was of great importance for the consolidation of the Standard Model of particle physics.

The Stanford Linear Collider (SLC) reached similar energies a couple of years later. This machine used the linear accelerator of SLAC to accelerate electrons as well as positrons to about 50 GeV. The two particle beams were deflected by 270° with a strong magnetic field to bring them to a collision.

Virgin territory was entered when the new accelerator HERA started to operate at DESY in Hamburg in the '90s. HERA produced head-on collisions between electrons with energy 30 GeV and protons with energy 800 GeV for the first time. With the help of the HERA machine it was possible to do an especially precise analysis of the inner structure of the proton, down to distances around 10^{-16} cm, which is a thousandth of the diameter of the proton.

Electrons and positrons have one disadvantage. They cannot be accelerated to very high energies in a ring accelerator uncless the ring has a certain size. This is because electrons and positrons have small masses. According to the theory of QED they can easily emit electromagnetic radiation. When the direction of their momentum changes, part of the particle's electromagnetic field flies off in the form of electromagnetic radiation. This emission is called synchrotron radiation.

THE FUNDAMENTAL CONSTANTS

Electrons flying in circles in a ring accelerator are constantly changing the direction of their momentum. So they act as a source of this radiation, and energy is lost constantly. The closer the electrons' velocity gets to the speed of light, the stronger the synchrotron radiation becomes. Finally, hardly any energy is left to accelerate the particles. This problem is avoided if the fast-moving electrons move in a straight line. It is therefore considerably more efficient, if one wants to obtain energies of several hundred GeV, to accelerate electrons and positrons in a linear accelerator rather than in a ring.

Consequently, there are plans to construct a big linear accelerator as a global particle physics project. This linear accelerator, called the International Linear Collider (ILC), would produce electrons and positrons with an energy from about 250 GeV to 1000 GeV. The ILC would be the ideal machine to test today's theory of particles and particle interactions under extreme conditions. It would presumably even discover physics effects which go beyond the Standard Model.

A decisive step was taken at the beginning of the '80s at CERN, when the first high energy proton–antiproton collisions were produced. Antiprotons do not exist in the matter around us, but they can be created in particle collisions and then accelerated just like protons. However there are technical difficulties in producing a "clean" beam with a large number of antiprotons. It took many years of intensive research to overcome this problem.

As in electron–positron machines, protons and antiprotons can fly in opposite directions through the same vacuum tube of a collider if both particle beams have the same energies. Proton–antiproton collisions studied at CERN, starting from 1981, finally reached energies of 400 GeV per particle.

Einstein: It must be very interesting to have matter colliding with antimatter — I'm sure there was a lot to be learned.

Haller: Yes, the investment paid off. The W and Z particles were soon discovered by a group of physicists led by Carlo Rubbia. These particles had been predicted in the theory of electroweak interactions proposed in the '60s by Sheldon Glashow, Abdus Salam and Steven Weinberg. They are responsible for the transmission of the weak force, which, for instance, causes radioactive decay.

Fermilab took the lead in the study of proton–antiproton collisions from the '90s. Fermilab's big ring was changed into a proton–antiproton collider. The difficulty that I mentioned before — that synchrotron radiation is produced when electrons travel in a circle — also exists for protons, at least in principle. The large mass of the proton, however, makes it unimportant for all practical purposes. That is why it is possible to accelerate protons or antiprotons in a ring to energies of many thousands of GeV without any considerable loss of energy through synchrotron radiation.

The new LHC at CERN will reach energies of 7000 GeV per beam using superconducting magnets with a field strength of 8.4 Tesla. Let me emphasize that the LHC ring is in the same tunnel that formerly sheltered the LEP accelerator.

The LHC will open the door to energies above 1000 GeV. Though there is no lack of theoretical speculation, it is not yet foreseeable what physics will look like at this high energy. But there is hardly a particle physicist who doesn't believe that the energy range encompassed by the LHC will bring us surprises.

Einstein: Well, we shall see. That certainly was an interesting presentation. Now I know where the problems in particle physics are and how they can be solved through experiments with accelerators. I think that we deserve a coffee in the Athenaeum.

The three physicists went to the restaurant and enjoyed a good Italian espresso. Einstein ordered ice cream again. This time he had lemon ice cream with lots of whipped cream.

COLORFUL QUARKS AND GLUONS

6

After the coffee break the three physicists reconvened in the library of the Athenaeum.

Einstein: Now let us return to the physics. The idea of the quarks makes good sense to me, but there is one thing that I don't understand. Atoms are well described by the quantum field theory of QED. Quarks must be described by a similar theory. But which one? Mr. Haller, is there something new?

Haller: Yes, we have now a theory that is very similar to quantum electrodynamics that can describe the forces between quarks. These forces are responsible, among other things, for the fact that three quarks always come together to form a nucleon. The theory is called quantum chromodynamics, or QCD. It was developed in 1972 by Murray Gell-Mann and his German collaborator Harald Fritzsch. Both were at CERN in Geneva at the time.

The name quantum chromodynamics is derived from the Greek word *chromos*, meaning color. Quarks, such as the u quark, have a property that is a new type of charge. This charge has nothing to do with the electric charge. Whereas the electric charge can be described by one number, say $+1$ or -1, the new charge is described by three numbers, and it is called the color of the quark. There are exactly three such numbers — the number three plays an important part in the theory. There are, for example, a red, a green, and a blue u quark. A red quark can be described by the numbers $(1, 0, 0)$, a green quark by the numbers $(0, 1, 0)$, etc.

Initially, Gell-Mann liked to use the French national colors red, white, and blue, since he lived in a house not far from the village of Gex, in the Jura Mountains near Geneva. But Fritzsch preferred the colors red, green, and blue, because white is not a primary color. When one mixes red, green, and blue, one obtains white, and that turned out to be very useful later.

Fritzsch and Gell-Mann, together with their American colleague Bill Bardeen who was also at CERN, studied the decay of the neutral pion. This particle decays very rapidly into two photons due to an electromagnetic process. The three physicists found that the strength of this process depends on the number of colors. If there are no colors, one finds that the decay rate of the pion is too small by a factor of nine. This argument was used at first to discredit the quark model. But if there are exactly three colors, the decay rate increases by a factor of nine — the square of the number of colors — giving an excellent match with experimental results. That is a nice success for the quark model if one takes the quark colors seriously.

The fact that there are three colors is important because it is responsible for nucleons being composed of three quarks. The assumption is that all three colors must be present in a nucleon. If we associate the quark colors with the three primary colors red, green, and blue, this means that physical particles are white, since a mixture of red, green, and blue gives white.

It should be emphasized that the symmetry of the colors should be an exact symmetry. In physics we mostly encounter broken symmetries, but in the case of the quark colors we are, exceptionally, dealing with an exact symmetry. Nobody knows why nature has arranged it this way.

Einstein: Well, I expect that God had his reasons. A symmetry that is not broken represents something beautiful, and that is rare in our world. I would like to mention another thing that is to God's credit: he arranged the symmetry of the colors in such a way that one does not even notice it. Heisenberg was not aware of the colors, and Gell-Mann and Fritzsch stumbled upon the symmetry only after a long process. God introduced the symmetry, but then made sure that it would not attract any attention. Only the very careful observer would notice it. That is excellent. God is never obtrusive, but often rather subtle. I must praise him for this.

Haller: Let's take another look at the old quark model. Before the discovery of quarks as particles inside nucleons, physicists had noticed a strange property of the quark model. Consider the naive quark model, in which nucleons are interpreted as a system of three quarks.

The excited states of the proton had been discovered during the scattering experiments in the '50s with the Bevatron in Berkeley. These states have spin $\frac{3}{2}$ and they decay into a nucleon and a π meson. We have already mentioned them — they are called delta particles. One was especially remarkable: the Δ^{++} particle, which has a mass of 1230 MeV. In the context of the naive quark model, this must be an object that is made up of three u quarks and has an electric charge of +2. If the spin of each of the three quarks points in the same direction, we have a system with the required spin $\frac{3}{2}$.

Einstein: Just a minute, this particle seems strange to me now. It is made up of three u quarks, and they have the same spin. Doesn't that mean that we have a problem with the Pauli principle, or do the colors of the quarks help here?

Haller: Congratulations, Mr. Einstein, that's right. Quarks are objects with spin $\frac{1}{2}$ and should be subject to the Pauli exclusion principle, as are all the other particles with spin $\frac{1}{2}$. This principle was discovered by Wolfgang Pauli soon after the advent of quantum mechanics. It says that for a compound system of spin $\frac{1}{2}$ objects, the state is antisymmetric if two particles are exchanged. This is not the case for the state we are discussing. It is symmetric, and so we have a problem.

We should give an example to elucidate the principle for Isaac Newton. Imagine a state made up of A and B, let's call it AB. If we exchange A and B, we obtain the state BA, which is different to AB.

If A and B were objects with spin $\frac{1}{2}$, according to the Pauli principle neither the first nor the second state could be present in nature, but the state $(AB - BA)$ can exist. This state changes its sign when A and B are exchanged — it is antisymmetric. The state $(AB + BA)$ is symmetric and would not be allowed by the Pauli principle. The Pauli principle describes a subtle property of quantum physics which is of great importance for atomic physics.

Einstein: The Pauli principle is mysterious to me. Can it be related to something else? Or can it be derived?

Haller: Yes, you will be astonished. The Pauli principle can be derived from your relativity theory, but the proof is very complicated and I don't suggest that we discuss it now.

Let's look instead at the application of the Pauli principle to quarks. The Δ^{++} particle has the quark structure (uuu). Exchanging two quarks reveals the problem: nothing changes. This means that the state should not exist in nature. This conflict between the simple quark model and reality was one of the reasons why the quark model in 1964 encountered such strong resistance from many physicists. As I've already mentioned, the solution to this problem, the color quantum number, was found by Fritzsch and Gell-Mann in the early '70s.

The colors give us new ways to construct a state from three u quarks. We could write $(u_r u_r u_r) = (rrr)$ for example. This would be a state made up of three red u quarks. But since this state is symmetric with respect to the exchange of two quarks, we still have a problem with the Pauli principle. There is only one state that is antisymmetric with respect to the exchange of two quarks, and that is the state (rgb − rbg + brg − bgr + gbr − grb). This state has the property that all colors are present, no color is underrepresented or overrepresented, and it is completely antisymmetric.

The symmetry associated with rotation of the colors has certain similarities to phase rotation in QED. Fritzsch and Gell-Mann suggested in 1972 that the color charge be treated in a similar manner to the electric charge in QED. The result is the gauge theory named QCD. The theory was met with extreme skepticism initially, but it became clear in the '70s and '80s that QCD is capable of describing the dynamics of quarks inside nucleons extremely well. Today it is recognized as a theory of the strong interaction.

Before we go into more detail, let me explain a fundamental difference between QED and QCD that has to do with the structure of the symmetry. The gauge symmetry of QED is very simple. We are dealing with the symmetry of a phase rotation. For example, we might rotate by 30° forward, then by 10° backward, and then by 37° forward, and so on. A rotation can always be described by a number, a parameter, that is the angle of the rotation. Executing two rotations one after the other, say first by 10° and then by 30°, gives us a rotation of 40° in total. If one first rotates first by 30° and then by 10°, one also ends up with a rotation of 40°. The result does not depend on the order of the transformations. To mathematicians, this kind of symmetry is known as an Abelian symmetry after the 19th century Norwegian mathematician Niels Henrik Abel.

The symmetry of the colors is not so simple. We can demonstrate this with an example from geometry. Rotations in a plane, i.e., in a two-dimensional space, are an example of an Abelian symmetry. If we now introduce a further dimension so that we have a three-dimensional space, there are three possible rotations that are independent of each other: rotations around the x-axis, the y-axis and the z-axis, also called the three Euler angles. Any rotation in space can be described using three parameters.

These rotations can be arbitrarily combined. We could, for example, first rotate by $10°$ around the x-axis, and then by $20°$ around the new y-axis. Or we could rotate first around the x-axis by $20°$, and then by $10°$ around the new y-axis. In both cases, we have a rotation in three-dimensional space. However, as is easily verified by an experiment or a small calculation, the two results are not identical. They differ by a rotation around the z-axis. The result depends on the order of the various rotations. This kind of symmetry is known as a non-Abelian symmetry.

A gauge theory that is based on a non-Abelian symmetry is called a non-Abelian gauge theory. The American theoreticians Chen Ning Yang (born in China) and Robert Mills were the first to study such theories in the 1950s. But the first ideas about constructing such theories had been contributed years earlier by Hermann Weyl in the United States, Oskar Klein in Sweden, and Wolfgang Pauli in Switzerland. Pauli went so far as to write a lengthy letter to Princeton about his ideas. It can't be excluded that Yang, who was at Princeton at that time, used Pauli's ideas in his work without emphasizing the fact. In any case, he never explicitly mentioned Pauli's letter as a source. Today the content of Pauli's letter is known, and it does contain essential elements of the non-Abelian gauge theory or the Yang–Mills theory. Pauli should have published his ideas. If he had done, we would speak today of the Pauli–Yang–Mills theory.

Yang and Mills wanted to apply the theory to the weak interactions, but they could not work out how to introduce a mass for the particles that carry the weak force, known as W bosons. It was not until much later that a way was found to incorporate a mass without causing problems in the theory. This is a difficult topic we will return to later.

In 1967 Fritzsch wrote his diploma thesis in the field of gravitation, at an institute of the Academy of Sciences in Potsdam in East Germany. He wanted to solve the problem of quantizing gravitation. His professor suggested taking a look at the Yang–Mills theory as preparation. This he did in great detail. So Fritzsch knew quite a bit about Yang–Mills theory, more than most other theoreticians at the time. Suddenly he found he could apply this knowledge to the quarks.

Einstein: That's good to hear. Someone wants to quantize gravitation, a hitherto unsolved problem, and ends up learning about Yang–Mills theory. I have one question: Fritzsch was in Potsdam, that's in East Germany, so was he locked up in the DDR? How could he have worked with Gell-Mann? That's pretty amazing for a physicist who lived in East Germany. It's a remarkable career, comparable to that of George Gamow, who came from Russia to the USA.

Haller: Fritzsch was politically active in the East, in opposition to the regime. He was involved in several activities that could each have gotten him a minimum of 10 years in jail, so in the summer of 1968 he and a friend escaped together to the West. They crossed the Black Sea from Varna in Bulgaria to Turkey in a small folding canoe. This escape across the open sea even interested the CIA, who interviewed them extensively afterwards. The CIA then offered to send the two physicists to any American university as doctoral students, at its cost. The CIA was apparently impressed by the two and wanted to lure them to the United States. But Fritzsch could work at the Max-Planck Institute in Munich with Werner Heisenberg, and his friend was invited to DESY in Hamburg, and they stayed in West Germany.

Einstein: That's what I call courage — crossing the Black Sea, which was full of Soviet warships, in a boat! I don't think I could do that. George Gamow tried it once as well, in the 1930s, in a sailboat starting from the Crimea. In the end, he was happy to be saved by a Soviet warship. In any case, Fritzsch and his friend made it, apparently without the help of Soviet warships.

Haller: Yes, the escape also impressed Gell-Mann greatly, and this was the real reason that he started to work with Fritzsch. Gell-Mann and Fritzsch constructed a Yang–Mills theory in the space of the colors of the quarks, calling the force-carriers gluons. The theory was introduced at the big Rochester conference on particle physics in Chicago in the fall of 1972, but most physicists didn't take it very seriously. Some joked: "Three quarks might be OK, but nine quarks, that's too much nonsense, and to add eight gluons on top of everything is just terrible." But Fritzsch and Gell-Mann continued to work on the theory, and a year later the theory began its march to success.

But now to some of the details of the theory. Compared to rotations in ordinary space, rotations in color space are more complicated. Since quarks are described in terms of complex fields, similar to the description of electrons in QED, the color symmetry is obtained when the three axes x, y, and z are replaced by complex axes. This symmetry is called an SU(3) symmetry in mathematical jargon. The number of parameters needed to arrive at any single transformation in color space is considerable, namely eight. This means that five more parameters are needed to describe a transformation in color space than are needed to describe a simple rotation in three-dimensional space. It is easy for a layman to understand how we arrive at eight parameters. Since there are three colors, we can characterize a transformation by stating which color turns into which color: $r \rightarrow r$, $r \rightarrow g$, $r \rightarrow b$, $g \rightarrow r$, $g \rightarrow g$, $g \rightarrow b$, $b \rightarrow r$, $b \rightarrow g$, $b \rightarrow b$.

Newton: But that's nine possibilities, not eight!

Haller: Yes, I've given you nine possibilities. The case where the color stays the same must be counted, as in QED, but there is a case that should not be counted. According to the theory, we do not need to count the case when all three colors stay the same, i.e., the superposition r → r + g → g + b → b. That leaves us with 9 − 1 = 8 possibilities. If the number of colors were two instead of three, we would have 4 − 1 = 3 possibilities. That is equivalent to the three normal rotations in a three-dimensional space. For four colors, there would be 16 − 1 = 15 possibilities, but luckily there are only three colors.

Gell-Mann liked to refer to the fact that the number eight plays such an important role in the SU(3) symmetry as "the eightfold way", an analogy to Buddhism. The Eightfold Path in Buddhism is composed of right view, right intention, right speech, right action, right support, right effort, right mindfulness, and right concentration.

If we allow that the color symmetry is a gauge symmetry and demand that the respective field equations respect this symmetry, we end up predicting an interaction between quarks and gluons analogous to the interaction between electrons and photons in QED. Photons transmit an electromagnetic force that acts on electric charge. Analogously, gluons transmit a force that acts on a quark's color charge. But due to the nature of the gauge theory in color space, this interaction is different to the one in QED.

In general, a non-Abelian gauge theory has a structure that differs considerably from an Abelian gauge theory such as QED. In particular, the coupling of the gluons to the quarks is different to the coupling of the photons to the electrons. An electron changes its momentum when interacting with a photon, but it remains an electron. A quark, on the other hand, can change its color state when interacting with a gluon. For example, a red quark can turn into a green quark. These interactions are given by the transformations in color space. Since there are eight different transformations, we have eight different gluons. These gluons are characterized by their color carrying properties. So there can be a red → green gluon or a green → blue gluon, etc.

We can use the analogy between QED and QCD to compile a little dictionary that connects the concepts of the two theories.

electron	quarks
QED	QCD
electric charge	color charge
photon	gluon
atom	nucleon

As previously mentioned, one of the most marked differences between QED and QCD is the number of quanta needed to transmit the force: one photon in QED and eight gluons in QCD. Another important difference is that the charge of a particle involved in an electromagnetic interaction does not change, but the color of a quark may change when it interacts with a gluon.

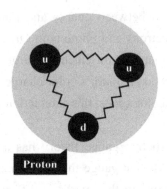

Three quarks form a nucleon, held together by gluons.

And there is yet another important difference. The fact that photons are electrically neutral and therefore cannot undergo any interactions with themselves is very important. It is because of this that the myriads of photons in a laser beam, for example, can fly at the speed of light side by side through space without influencing each other. Gluons could not do this. It turns out that gluons not only undergo interactions with quarks, but also with themselves. Gluons have a color charge just like quarks.

The fact that gluons also have color, that they are charged in the sense of color symmetry, has dramatic consequences for the interaction of quarks and gluons. It changes the vacuum polarization. According to QCD, the vacuum is filled with virtual quarks, antiquarks, and gluons. If we observe the region around a quark, we find that it is affected by the quark's chromodynamic interactions. As a consequence of its color charge, the quark pushes the virtual quarks out of its immediate vicinity, while at the same time attracting antiquarks. You might expect this to produce a similar effect to that familiar from our discussion of quantum electrodynamics, i.e., that the effective color charge of the quark would be partially shielded. But since gluons carry a color charge, the sea of gluons surrounding the quark is also polarized. This does not happen in QED. An electron is surrounded by virtual photons, but since photons carry no charge, they are not influenced by the other photons.

When the vacuum polarization effect in the non-Abelian gauge theory was examined closely in the '70s, the effect, surprisingly, was found to be qualitatively different to that in QED. The initial calculations were performed by the Russian theoretician Iosif Khriplovich and the Dutch physicist Gerard 't Hooft, and later by David Gross and his student Frank Wilczek, as well as by David Politzer at Harvard. At that time Politzer was a doctoral student of the famous physics professor Sidney Coleman in Harvard, who also contributed to the calculation. To everyone's surprise it was discovered that the virtual gluons cause an increase in the color charge at large distances. This has a direct influence on the strength of the interaction in QCD, characterized by a constant analogous to the fine structure constant α of QED. This constant is generally known as α_s.

As I've already mentioned, the electromagnetic fine structure constant becomes larger at very small distances. It is exactly the opposite in QCD. The polarization of the vacuum due to gluons reduces the parameter α_s at small distances. Strictly speaking, α_s is not really a constant, but a function of the distance, or energy, involved in an interaction.

The interaction parameter α_s has been measured in many experiments, in electron–proton reactions, for instance, and in electron–positron annihilations.

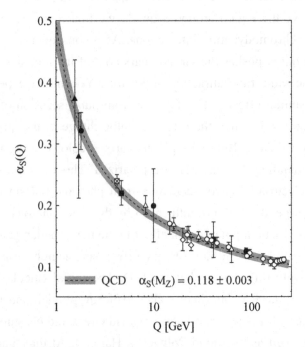

The behavior of the chromodynamic coupling constant as a function of energy.

As an example, let us take the value of α_s at an energy of 91 GeV. That is the mass of the Z particle, which we will discuss later. One finds $\alpha_s \approx 0.12$.

The interactions in QCD are therefore stronger at this energy than they are in QED. But since α_s is still small compared to 1, it is possible to use methods similar to those used in QED, based on perturbation theory, to calculate the details of the quark–gluon interaction. However, α_s becomes larger at smaller energies. Correspondingly the interaction becomes stronger at smaller energies, equivalent to a greater separation of the particles. Finally perturbation theory is no longer applicable.

Calculations show that the mass of the quarks becomes infinite. In QCD, that leads us to the conclusion that quarks cannot exist as free physical particles, they are permanently tied up in nucleons. This is the current opinion, at any rate, though not everything has been proven.

Einstein: I like this theory. It is interesting, almost mysterious, that the fundamental particles cannot be free. It is a beautiful theory and because it is beautiful, it is probably true — my compliments to Fritzsch and Gell-Mann.

Haller: Yes, we believe that the theory is right and beautiful, in some ways simpler and more beautiful than your theory of gravitation.

Einstein: Well, well... more beautiful than my theory of gravitation? I do not agree. But you have to remember in my favor that I invented the theory of gravitation, and the inventor has a right to be proud of his invention.

Haller: OK, we don't want to argue about this. To me, both theories are beautiful, and I think that both theories are right.

It is instructive to examine the interaction between a quark and its antiquark. At small distances — small compared to the characteristic scale of strong interactions, around 10^{-13} cm — the force between them behaves like an electric force, decreasing as the square of the distance. In this regime, we can describe the chromodynamic force as we do the electric force, with field lines.

When the distance becomes greater than 10^{-13} cm, the situation changes. Since gluons interact with each other, there is effectively a force acting between the chromodynamic field lines that pulls them together. We can compare this effect to a phenomenon in electrodynamics: two parallel wires that carry currents flowing in the same direction will attract each other. This effect is a result of the magnetic field that surrounds each wire. In QCD the interaction of gluons produces an attractive force between the gluonic field lines.

While the electric force between a quark and an antiquark decreases with increasing distance, the chromodynamic force does not. Thus it is impossible to separate the quarks. The situation has been examined in a model where space and time are not continuous, but discrete — in a lattice gauge theory. In this model, when quarks are pulled very far apart, the field lines look completely different to those in electrodynamics. The field lines emanating from one quark do flow to the other quark, but they are almost parallel to each other. The field lines resemble a hose, similar to the electric field lines between the two plates of a parallel plate condenser. The force between quarks is also similar to the force between the plates of a plate condenser; it is constant, independent of distance.

Newton: But you said that in this model time and space are discrete, so they can only take on certain values. Does something like that make sense? Time and space are continuous phenomena after all.

Haller: Yes, it's just a calculation trick. What we are using here is called a lattice space. It has the nice property that we are capable of doing calculations in it that would be impossible in continuous space. If we do the calculations in this space, we have exactly the situation that I have just described. The field lines become parallel, and the quarks are trapped forever. If we consider making the distance between two lattice points smaller and smaller, we can infer what it would look like if space were continuous. So working with a lattice space does not mean that we are robbing space and time of its continuous character. The point is simply to be able to do calculations that would otherwise be impossible.

On the other hand, the force acting between quarks becomes very weak at very small distances. This explains why quarks behave like structureless, unbound objects in electron–proton scattering experiments. In these experiments, the electron, which is not influenced by QCD forces, penetrates deeply into the proton and interacts with one quark very briefly, making the gluonic interaction hardly noticeable.

A comparison with everyday life will help explain this: imagine three colored balls, one red, one green, and one blue, moving fast and bouncing off the walls inside a hollow glass sphere. Now let the average time between bounces be only a hundredth of a second. If we photograph the sphere, choosing a relatively long exposure time, say a fifth of a second, we will see nothing of the balls. What we see is a white sphere, since the colors of the balls average out, leaving a white picture. If, however, we choose an exposure time of a thousandth of a second, we will see all three balls and their colors very clearly.

This situation is very similar to that with quarks in particle physics. When we perform high energy scattering experiments with electrons, we see the three quarks in a nucleon very clearly. This is because the exposure time, given by the duration of the collision, is very short. When we use low energy electrons, we see nothing or very little of the quarks. So with a short exposure time the quarks appear to be free particles, and with a long exposure time they appear as strongly bound objects.

The decrease of gluonic forces for small intervals of time and space is known to physicists as asymptotic freedom. At very small distances, the quarks are free particles. One consequence of this phenomenon is that the interaction constant α_s becomes smaller at smaller distances or with growing energy scale. This makes it possible to apply the methods of perturbation theory as in QED.

Newton: That's really interesting. Suddenly we have the possibility of using perturbation theory even for the strong interaction.

Haller: Yes, suddenly it works, and that was a surprise. Nobody had been expecting that. Furthermore, QCD has made an interesting contribution towards today's understanding of the strong nuclear force inside the atomic nucleus. The nuclei are made up of nucleons, and these are in turn composed of quarks. The motivation at the beginning of particle physics was to understand the nuclear forces better. What forces, for instance, are responsible for assembling six protons and six neutrons in one of the most stable atomic nuclei, the carbon nucleus?

Newton: Maybe these forces have something to do with the color forces?

Haller: Yes, that's how it is. Today we can say that these forces are by no means fundamental, but are rather indirect consequences of the gluonic forces between the quarks inside the nucleons. The forces between electrically neutral atoms responsible for the formation of molecules are comparable; those are indirect consequences of the electric forces inside the atoms.

Today, now that the secrets of the atomic nuclei and of the nucleons have been lifted, we have a picture of the nuclear forces and their building blocks. The nuclear forces can be understood in the framework of quantum field theory. Even though nature only needs the two quarks u and d to form stable atomic nuclei, there are six quarks at work in the concert of subnuclear forces. An atomic nucleus seems calm on the surface but deep inside there is the bubbling complex microcosm of unstable hadrons, as can be seen in the high energy collisions.

Einstein: Good. Contrary to the opinions of my younger colleagues, including Werner Heisenberg, I always thought that the nuclear forces were not directly fundamental. It seems that I was right in the end.

Haller: Yes, you were absolutely right about the nuclear forces. But now that we are reaching the end of our discussion of QCD, we should see what the consequences of this theory are for the structure of objects that are composed of quarks.

Quarks, like gluons, do not appear as free particles since they have color. There are exactly three colors, and so the quarks are said to form a color triplet. All three colors are equal and interchangeable. This leads to a color symmetry that is described by the group SU(3). We went through the arguments why colored quarks cannot exist as free particles.

The same reasoning applies for all colored objects in QCD, including the eight gluons referred to as a color octet. The only objects that can appear as real particles are those that show no color to the outside. In the language of mathematics, these objects, in which the colors of separate building blocks cancel, are called color singlets.

This effect is well known in QED. Atoms are objects made up of electrically charged building blocks — the nucleus and the electrons in the shell — but the atoms themselves are neutral. In keeping with the analogy, we could refer to them as charge singlets.

We have not yet examined the simplest color singlets that can be made from colored quarks. A quark and an antiquark cancel their color charges exactly, thereby forming a color singlet. These objects, made half of matter, half of antimatter, exist in nature as unstable particles. They are produced in particle collisions and decay soon after. We mentioned them earlier; they are called mesons. The first meson was discovered in 1947 in a cosmic ray experiment. It was an electrically charged particle with no spin and a mass 207 times that of an electron, or about 140 MeV. That made this particle, called the π meson, considerably lighter than a proton.

We can also use only quarks to construct color-neutral objects. Because we have three colors, it is possible to construct a color-neutral object, for example a nucleon, out of three quarks. In electrodynamics, which can be viewed as a gauge theory with one color charge, this would not be possible. Neutral states, such as an atom, can only be constructed in electrodynamics by compensating every positive electric charge with a negative one.

Einstein: All the particles that you have mentioned so far are made up of quarks, while the gluons play a subordinate role as the particles that transmit the force between the quarks. But two gluons, which are color octets, could easily be combined to form a singlet. Then we would have particles that are color neutral, made up of only gluons. Do such particles exist?

Haller: That is an interesting but difficult question. Fritzsch and Gell-Mann introduced such particles and called them glue mesons, although some physicists speak of glue balls, as if the objects were tennis balls. Experimental physicists have been looking for such particles without definitive success so far. But this is a difficult problem. Glue mesons can mix with normal, electrically neutral mesons, and then things are not so clear. A number of neutral mesons have been discovered, but it seems to be impossible to decide for sure if they are glue mesons. They are probably mixed states rather than pure states, consisting of quarks and antiquarks part of the time and of gluons the rest of the time. The experimentalists keep looking.

Towards the end of the '70s, with the help of SLAC's high energy electrons, it was possible to obtain a kind of X-ray picture of a nucleon. The pictures mainly show that there are in the nucleon three electrically charged objects without structure and with charges $+\frac{2}{3}$ and $-\frac{1}{3}$, the quarks.

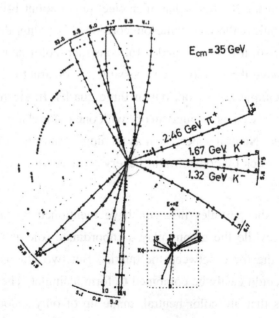

Gluon jets as seen at DESY, Hamburg.

At high energies, quarks exist as practically point-like structures, and we can disregard the strong interactions. At low energies, quarks are under the influence of very strong forces. The disappearance of the forces among quarks at high energies is called asymptotic freedom. At low energies, the opposite happens, and this is referred to as "infrared slavery".

Einstein: I do not like the word "gluon", in which the English and Greek language mix. Fritzsch and Gell-Mann should have chosen a more appealing name. Chromon, perhaps? Gluons are, after all, very important, and they deserve a better name than "gluon".

Haller: You are right about that. I didn't like the name, and neither did Fritzsch. Actually, he did suggest the name "chromon", but Gell-Mann had introduced the term "gluon" in his first paper on quarks, and he insisted that the name be kept.

The first indirect evidence of the existence of gluons was found at the end of the 1970s at the research center DESY in Hamburg. There scientists studied processes that would every now and then emit a gluon at high energy. The gluon revealed itself as a jet of particles which could be observed. This jet had been predicted by theory, and it appeared in addition to two jets associated with quarks. So, although quarks and gluons are permanently confined, we can see them indirectly as jets of particles, most of which are mesons. If a quark is emitted in a collision with high energy, quark–antiquark pairs are created between the quark and the other particles. These pairs form mesons. Thus we see a jet of mesons, which describes the quark. The energy or the momentum of the jet is essentially the energy or the momentum of the quark. This is called a quark jet. If a high energy gluon is emitted in a collision, one obtains again a jet of mesons, which is called a gluon jet.

Einstein: Electrodynamics and chromodynamics have one thing in common. In both theories, electrons and quarks are point-like objects with no inner structure. But are they truly elementary? Have the experimentalists tried to measure the size of leptons and quarks?

Haller: Yes, people have looked for an inner structure of the leptons and quarks, so far without success. A limit on the possible radius of quarks was set with the help of the big accelerators LEP at CERN in Geneva, the HERA machine at Hamburg's DESY, and the Tevatron accelerator at Fermilab near Chicago. This limit is around 10^{-17} cm, or about $\frac{1}{2000}$ of the nuclear radius.

But looking at my watch, I see that it is already quite late. We have just enough time for a little coffee break. I suggest we go to a café I know on Lake Street.

THE STANDARD MODEL

7

About an hour later the physicists were back in the library.

Haller: In our world we find, besides the electron, five other particles that are seemingly without structure and that do not take part in the strong interactions. These particles are called leptons. The electron is the lightest electrically charged lepton. It is accompanied by the muon, which we have already mentioned. The muon's exact mass is 105.658 MeV; it is about 207 times heavier than the electron.

Einstein: Remember that you mentioned how Isidor Rabi always made fun of this and asked, ironically, "the muon, who ordered that?" It seems to me that you still don't have an answer, not even the faintest idea of one.

Haller: No, it is still a mystery.

Einstein: Mr. Haller, you're making me curious. There should be a reason for the existence of the muon. Do we really not know anything?

Haller: I'm telling you, no one has the slightest idea. We could all live well without the damn muon, but the muon exists. We have no idea why, but we have stopped asking.

Nature may have its reasons for the muon. Or maybe God just made a stupid mistake and introduced it by accident. But the muon is useful, at least, for measuring the interactions of particles. Experimental physicists love muons and study them in their experiments. Most particle physics experiments are surrounded by muon chambers, and these chambers tell us quite a lot. We also have a particle called the tauon, with a mass 3540 times the mass of an electron, or 1776 MeV. This number 3540, dear Mr. Newton, is another one of the many natural constants that we don't understand.

Both particles, the muon and the tauon, are unstable and decay soon after their creation in particle collisions. The muon has its uses, as I've just mentioned, in particle physics experiments. But we could really have done without the tauon. We know of no application where it could be of any use. Only Martin Perl from SLAC has profited from this particle; he received the Nobel Prize for the discovery of the tauon.

The three electrically charged leptons are associated with three electrically neutral particles, the neutrinos. To every charged lepton, there belongs a neutrino particle.

Einstein: My friend Wolfgang Pauli introduced the neutrinos at the end of the 1920s. He was trying to save the law of energy conservation, which had seemingly been violated by radioactive decay. I noted his hypothesis, but I did not take it very seriously. So you say that neutrinos really exist? Are you sure?

Haller: Yes, absolutely sure. They were discovered in an experiment carried out by the Americans Clyde Cowan and Frederick Reines in a hole in the ground near an atomic reactor, the Savannah River reactor in Ohio, belonging to the US army. Curiously, Cowan and Reines were only allowed to publish incomplete results. For instance, they were not allowed to state the distance of the detector from the reactor's core. This was because such information would have enabled the Russians to calculate the power of the reactor. From this, they could have deduced how many atomic bombs it could produce, and that was a secret. In any case, Cowan and Reines found sufficient evidence that a lot of neutrinos were being emitted by the reactor.

The experiment was carried out before the mid-1950s, so Pauli had a chance to see its success. When Pauli had introduced neutrinos, he had said that he did so reluctantly because these particles would never be found. But he was wrong. Today we can do many things with neutrinos. In accelerator laboratories, intense neutrino beams are created and used in various experiments.

For a long time, we thought that neutrinos were massless, like the photon. But today we know that they have masses, albeit very small masses.

Einstein: Pauli always believed that neutrinos have a small mass.

Haller: Yes, but because no mass had been measured, it was really believed for a long time that neutrinos have no mass. I must confess that I always thought they would have a small mass, somewhere around a couple of electron volts.

An interesting effect related to the neutrino masses is the neutrino oscillations. These oscillations arise when there is a mixing among the neutrinos. A neutrino emitted in a beta decay, for example, does not necessarily need to be a mass eigenstate; it could be a mixture of two or three eigenstates. When the neutrino moves through space, the different states have slightly different speeds, depending on their mass. The composition of the neutrino therefore changes. A muon neutrino can become an electron neutrino, for example, and vice versa. This is called a neutrino oscillation.

In the 1970s I often gave lectures about neutrino oscillations, and I tried to convince the experimentalists to do experiments along these lines.

I was able to convince Professor Mößbauer in Munich to do a whole series of experiments — first at the reactor in Grenoble in France, and later at the reactor near Gösgen in Switzerland. But the experiments then showed no effects. This has changed since 2001. Various experiments have now shown that neutrino oscillations exist. The neutrinos studied today come free of charge — they come from the upper atmosphere, a sort of cosmic neutrino radiation. They have been studied with special detectors in Kamioka, Japan, and with the neutrino detector SNO in Sudbury, Canada.

The experiments force us to conclude that the neutrinos produced in radioactive decays are mixtures of mass eigenstates, with mixing angles that are relatively large. The experiments indicate that the mixing angles are of the order of 45°. But the neutrino masses must be very small. We do not know the exact scale yet, but there are indications that the neutrino masses should be much less than one electron volt. In future experiments it will be possible to determine the neutrino masses more exactly. Neutrino physics has therefore become a very interesting chapter of particle physics.

But let us now look at the quarks. You will recall that atomic nuclei consist of the two quarks u and d. The proton has the structure (uud) and the neutron (ddu). There are, however, four more types of quark denoted by the symbols s (strange), c (charm), b (bottom), and t (top).

We have already mentioned the s quark, which is important for the SU(3) symmetry. The four new quarks, which can be written as the pairs (c, s) and (t, b), are the building blocks of heavy particles. Like the muon, they are unstable. There are no stable particles made up of these quarks.

Einstein: Rabi could just as well have asked: "muon, tauon, s, c, b, and t quarks, who ordered this indigestible menu?" These are really spooky objects, totally useless. Nobody needs them.

Haller: You're asking too much of me. No one knows why these other quarks exist. Just like you, I could live quite well without them. The u and d quarks are quite enough, we don't need more.

Newton: You mentioned that the new quarks come in the pairs (c, s) and (t, b). The lightest quarks can also be written as a pair: (u, d). The six leptons come in three pairs, each pair consisting of a neutrino and a charged lepton. So I could write down three lepton–quark systems: (neutrino, electron, u, d), (neutrino, muon, c, s), and (neutrino, tauon, t, b). Would this make any sense?

Haller: Yes, we say that leptons and quarks come in three families, or three generations. But it remains unclear whether these families have a profound role in our theoretical understanding. Later we shall come back to this family structure, when we discuss the idea of Grand Unification.

Newton: What are the masses of the new quarks?

Haller: The masses are a puzzle. While the masses of the u and d quark are almost zero, the t quark is almost as heavy as the nucleus of gold (usually made up of 197 nucleons); its mass is about 174 GeV. So the t quark is a veritable monster. If we were to replace the u quarks in our body with t quarks, the mass of a person would increase from about 80 kg to around 25 tons. An elephant would be a ballerina in comparison.

Einstein: Oh God, that is giant, at least as far as mass goes. The size of the person would not change, though, since the sizes of the atoms are not changed. What is the mass of the s quark?

Haller: The s quark is relatively light. Its mass is only 150 MeV. But that already makes the s quarks 20 times heavier than the u or d quarks. The c quarks are even heavier, at about 1100 MeV, and the b quarks have a rather high mass of about 4300 MeV.

Newton: The new quarks will probably form particles with the old quarks. What are the masses of these particles?

Haller: That is easy to answer. First of all, we have the K mesons such as the negatively charged kaons, which are made up of an s quark and an anti-u quark. We also have neutral K mesons made up of one s quark and an anti-d quark. The charged and neutral K mesons have a mass of about 500 MeV.

There are also particles analogous to the proton, which have an s quark on top of a u quark and a d quark. So, for example, there is a particle made up of a u quark, a d quark, and an s quark, forming a color singlet. This particle, the neutral lambda (Λ) particle, was discovered at the beginning of the '50s in cosmic rays. I find it a beautiful particle because it unites all three light quarks.

In addition to the Λ particle, a number of other particles contain s quarks. Some have two s quarks, (uss) for example, and one, which has a charge of -1, is made up of three s quarks. This particle is called omega (Ω) and it has a mass of about 1670 MeV. It is the s quark analogue of the delta resonance (uuu). The difference is that it has an unusually long life time. The s quarks must turn into u or d quarks for the Ω particle to decay, and this is only possible through the weak interactions.

Because the existence of the Ω particle, and even its mass, had been predicted by Gell-Mann and Ne'eman using the SU(3) symmetry, the discovery of this particle at the American Brookhaven Lab in 1964 was a milestone in the development of 20th century particle physics. It cleared the way for Gell-Mann's Nobel Prize, which he received in 1969.

The charm quark together with an anti-u quark or an anti-d quark forms a particle called the charmed meson D, which has a mass of 1867 MeV. Analogously, there exists a B meson that is made up of a b quark and one of the light antiquarks; this particle has a mass of 5279 MeV.

Einstein: The new particles that contain t quarks must be interesting. They must have a huge mass.

Haller: In principle you are right, but nature plays a trick on us here. There are no particles made up of t quarks.

Einstein: Why not? A t quark and an anti-u quark could form a very heavy particle, almost as heavy as a gold nucleus. That particle should exist, at least for a short time.

Haller: Yes, in principle, but you must remember that t quarks decay into b quarks and a lepton pair or a pair of light quarks due to the weak interactions. This decay takes place very rapidly because of the large mass of the t quark; it happens so rapidly that no particle can form.

Einstein: Too bad, but back to the other leptons and quarks. I assume there is some kind of ordering principle — what is it?

Haller: We do not know much about it. The six leptons and quarks can be ordered in three pairs. It is even advantageous to put a lepton pair and a quark pair together into a lepton–quark family, as discussed before.

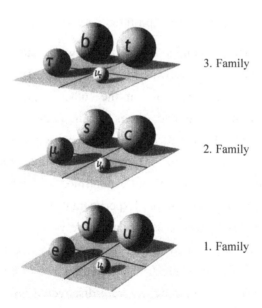

3. Family

2. Family

1. Family

The three families of leptons and quarks. A pair of leptons and a pair of quarks are combined to form a lepton–quark family. The particles of the first family are the u and d quark, the electron, and its neutrino. The second family contains the c and s quark, the muon, and its neutrino. The third family consists of the t and b quark, the tauon, and its neutrino.

The particles of the first family — the u and d quark, the electron, and its neutrino — are constituents of normal matter, except for the neutrino. The second family contains the c and s quark, the muon, and its neutrino. The third family consists of the t and b quark, the tauon, and its neutrino.

There is a more profound reason to group the leptons and quarks together in pairs. Asides from the already mentioned electromagnetic and strong interactions, there are also the weak interactions, which are responsible for the radioactive decay of atomic nuclei. The weak interactions, like the other forces, are mediated by the exchange of particles, here the W and Z particles. In contrast to photons and gluons, however, these particles have rather large masses, about 80 GeV for a W particle and about 90 GeV for a Z particle. These particles were found in experiments at CERN in the first half of the 1980s.

Newton: What is the theory behind these new massive particles?

Haller: The theory is a unified theory of the weak and electromagnetic interaction, called the electroweak interaction. The idea of the electroweak interaction came up in the '60s, in papers by Sheldon Glashow, Steven Weinberg, and Abdus Salam. They received the Nobel Prize for it in 1979. In this theory there were interactions that did not change the flavor of the leptons or quarks. For example, a neutrino could collide with an electron and remain a neutrino. Such interactions were called neutral current interactions. Experimental physicists claimed that such interactions could not exist, but they were wrong.

The theory we have today has a positive and a negative W boson, both with a mass of 80.40 GeV, and a neutral Z particle with a mass of 91.20 GeV. The W and Z boson were discovered around the end of 1983 at CERN with the help of the proton–antiproton collider. These bosons could not be directly observed. One can only observe their decay products. The W particles decay either into quarks, which cannot be observed directly, or, for example, into a muon and its neutrino. The neutrino cannot be seen, making the determination of the mass, at least, imprecise.

All reactions involving the weak interaction can be divided into two categories. First of all there are reactions where the electrical charge of the particle changes, e.g., in beta decay. In this case a neutron changes into a charged proton. These reactions are transmitted by W bosons. But there are also reactions where the electrical charge is not changed, for example, when neutrinos are scattered, as in the reaction

$$\nu + p \rightarrow \nu + p.$$

These reactions are transmitted by the Z boson. The first kind of interaction is called a charged current process, the second a neutral current process.

Neutral current reactions were controversial for a long time. Since reactions where an s quark is turned into a d quark were not observed, it was originally assumed that neutral current interactions did not exist. This was a mistake that misled the experimental physicists. The neutral current reactions were finally discovered in 1972.

Typical of all these different processes is that they involve four fermions. Sometimes a fermion is turned into three other fermions, as in the case of beta decay; sometimes two fermions react with each other, and in this case two other fermions can come out. This is the case in the reaction

$$\nu_\mu + n \rightarrow \mu^- + p.$$

I would like to emphasize that there is a considerable difference between electromagnetic and weak interactions with respect to the parity transformation, i.e., the transformation where all three space directions are reversed.

Let us examine an arbitrary fermion, an electron or a quark, for example. We can construct this fermion out of a left-handed and a right-handed fermion. A right-handed fermion is a particle that turns like a right-handed screw. A left-handed fermion, analogously, is a particle that turns like a left-handed screw. In a parity transformation, a left-handed fermion is turned into a right-handed fermion, and the other way around. If nature were symmetric with respect to parity transformations, then a left-handed fermion would have the same interactions as a right-handed fermion. This is the case for electromagnetic interactions but not for weak interactions.

Einstein: Oh my God, a violation of parity! Thus space is no longer symmetric under reflections?

Haller: Yes, I don't really like it, but that is how it is. The weak interaction violates the symmetry of parity.

This was observed in 1956, first in the radioactive decay of cobalt into nickel. The effect had been suggested by Chinese theoreticians living in the United States — Chen Ning Yang (from the Yang–Mills theory) and Tsung Dao Lee — and was experimentally verified in the same year. Lee and Yang immediately received the Nobel Prize.

Afterwards, it was found by Richard Feynman and Murray Gell-Mann that the violation of parity for leptons and quarks is very easy to describe. Only the left-handed leptons and quarks interact with W bosons, the right-handed ones don't. That makes the structure of the interactions between the W bosons and fermions essentially different from the structure of the interactions between fermions and photons. In the electromagnetic interaction there is no difference between the interactions of left-handed and right-handed fermions.

Why nature shows this bias towards left-handedness in the weak interactions is not understood. It is simply a fact that we must accept.

Einstein: That amazes me. I would never have thought that nature would be so weird, that it would be biased towards left-handedness. So nature loves the left, and the right is neglected.

Haller: Yes, but let us look at weak decays. Beta decay is a process in which a d quark emits a u quark and a negatively charged W boson. The boson, however, is virtual, and it immediately produces an electron and a neutrino.

Newton: The decay of a neutron into a proton still seems strange to me. Why does this happen at all? Why can't it happen the other way around? Can a neutron decay into a proton?

Haller: No, because the neutron mass is slightly larger than the proton mass. We know today that quarks have a small mass. And it turns out that d quarks are a little heavier than u quarks.

Einstein: Wait a minute — if quarks can't exist as free particles, why do they have a mass? I had this question in mind when you mentioned the masses of the quarks, but I didn't want to interrupt you.

Haller: Protons and neutrons would have the same mass if we neglected electromagnetic interactions and regarded the quarks that make up these particles as massless. In nature, however, we find that neutrons are a little heavier than protons, and this can only be explained if the d quarks are a little heavier than the u quarks. The quark masses are not really the masses of quark particles, but rather formal mass terms, needed in the QCD theory.

Thus the d quarks are a little heavier than the u quarks. Neutrons contain two d quarks and only one u quark. Thus neutrons are heavier than protons. A d quark decays into a u quark, and not the other way around. I hope that answers your question, Mr. Newton.

Even today we do not understand why the d quark is heavier than the u quark. It would be easier to understand if it were the other way around, since the charge of a u quark is almost double that of a d quark. For the other quark pairs, the mass of the quark with charge $+\frac{2}{3}$ is larger than the mass of the quark with charge $-\frac{1}{3}$: c is heavier than s, and t is heavier than b. But d being heavier than u has its advantages. Imagine if the u quark had a larger mass than the d quark...

Einstein: If that were the case, protons would decay into neutrons, Mr. Newton, and not the other way around. Then hydrogen could not exist, and, of course, there would be no water either. We would not exist, as we are 75% water.

Haller: Yes, that is a good reason for the d quark being heavier than the u quark: our existence depends on it! Thank God that it is the way it is.

But now let me continue. The electromagnetic and weak forces can be brought together in a unified theory called the theory of the electroweak interaction — the weak and the electromagnetic interactions are unified. The theory was first discussed by Sheldon Glashow, Abdus Salam and Steven Weinberg in the '60s. The theory of the electroweak interaction seems strange because the photon, the force carrier of the electromagnetic force, is massless, while the W boson has a large mass.

I have to admit that when I first heard about the theory of the two interactions, at the end of the 1960s, I was very skeptical, especially since the theory was formulated rather oddly. At the time, I was giving a seminar at the Max-Planck Institute in Munich. What disturbed me most was that the photon, first discussed by Einstein in 1905, was turned into a hybrid object in this theory.

The unification of the two interactions, the weak and the electromagnetic, comes about in this theory as follows. There are two neutral force particles in the theory, which have a certain mass. Furthermore there is a term that mixes the two particles. If we calculate the properties of the two physical particles, that is those which have a definite mass, one finds that one particle, which is today known as the Z boson, has a mass that is somewhat larger than the W mass; the other particle becomes massless, and that is the photon. The masslessness of the photon in the theory seems to be accidental.

Einstein: Yes, the theory is bizarre. The photon could easily have a mass. It is very strange that the photon should be massless by accident.

Haller: The difference between the mass of the W particle and the mass of the Z particle depends on a free parameter of the theory that is generally known as the weak mixing angle.

Einstein: So the photon turns out to be a hybrid. What a strange theory! And you believe that this theory is correct?

Haller: It looks like the theory is right. In any case, Glashow, Salam and Weinberg have already received the Nobel Prize for it. There are numerous experiments that confirm the theory; today nobody really doubts it anymore.

The first experiments were conducted in 1983 at CERN with the new proton–antiproton collider. Towards the end of 1983, W particles were observed, and soon thereafter the Z particles. The measured masses agreed with the theoretically predicted masses.

In 1989 a new collider called LEP started operation at CERN. In this accelerator electrons and positrons collided with each other.

When LEP started, I was convinced that small deviations from the electroweak model would be found. I did not believe in the theory, but I was wrong. Over the years, no deviations were found. The theory worked very well.

The same was true at the proton accelerator Tevatron at Fermilab near Chicago. Even with the help of the Tevatron, no deviation from the theory was found. The theory was perfect. Nowadays there is no doubt that the theory of the electroweak interaction is correct. Deviations could still exist, but they would have to be very small.

Einstein: That sounds impressive. Thus we have now a real theory of the strong interaction, QCD, and a theory of the electromagnetic and weak interactions, the electroweak theory.

Haller: Yes, today's Standard Model of elementary particles, which consists of the theory of the electroweak interaction plus QCD, seems to describe the whole of physics. This is an impressive result. With this model we can describe the fundamental forces and particles in the universe.

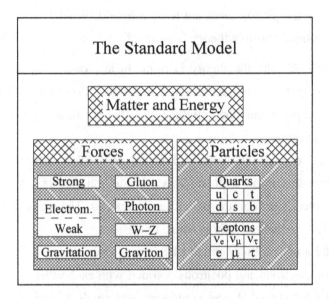

The Standard Model of particle physics.

However, this does not mean the end of research in fundamental physics. There are a number of important questions not answered by this theory. The most important one is how the particle masses are generated. The masses that are essential for the structure of the observed matter are the electron mass (0.511 MeV) and the proton mass (938 MeV).

Today we can say that the origin of the proton mass can be explained in the framework of QCD. The theory gives us a very clear and appealing picture. It says that the mass is nothing other than the field energy E of the quarks and gluons inside the proton. It is given by Einstein's relation $M = E/c^2$.

The proton mass and the masses of atomic nuclei represent the field energy of the quarks and gluons in the nucleons. What is important for this interpretation is the substructure of the nucleon. The mass and the internal structure of the proton are directly connected.

For leptons and quarks, it is a different story. In order to introduce their masses, physicists invented a hypothetical field which is able to give masses to these particles. This comes about through an interaction of the leptons and quarks with this field. The field is called the Higgs field after the Scottish theoretician Peter Higgs, who was one of the theoreticians who introduced this field. The masses of the W and Z particles are also generated by the Higgs field. But I should emphasize that in this model no masses can be calculated. The unknown masses, such as the mass of the electron, are replaced by unknown parameters describing the interaction of the Higgs field with fermions.

The Higgs field has a special property. It is a field that takes on a nonzero value in the normal vacuum. This is called the vacuum expectation value and is caused by the spontaneous breaking of a symmetry. To understand this, think of a Mexican hat. An apple lying exactly in the center, in the middle of the hat, will be elevated. The smallest disturbance, however, will cause it to roll down and come to rest somewhere near the edge of the hat. The apple now has a certain distance from the center. The rotational symmetry that was present when

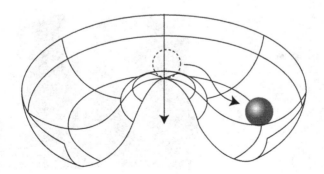

Model of the Higgs field, analogous to a Mexican hat.

the apple was lying in the center is destroyed. Physicists speak of a spontaneous breaking of this symmetry. The distance of the apple to the center signifies a certain energy. This is called the vacuum expectation value, and it generates the masses of the W and Z particles.

Like every other physical field, the Higgs field is connected to the existence of a particle, called the Higgs particle, which could be produced by particle collisions at sufficiently high energies. The necessary energies are so high that only the LHC accelerator at CERN will be able to produce the Higgs particle, if it exists.

I believe that the masses are generated through a mechanism other than the Higgs mechanism and that there is no Higgs particle. But even if this field exists, it is not clear whether it would produce not only the masses of the W and Z bosons but also the mass of the electron and of the other leptons. But maybe I'm wrong.

In the early 1970s, Gerard 't Hooft and Martinus Veltman from Utrecht showed that the theory of the electroweak interaction becomes renormalizable, i.e., calculable, if the masses of the W and Z bosons are introduced via the Higgs mechanism. They received the Nobel Prize in physics in 1999.

Gerard 't Hooft.   *Martinus Veltman.*

The masses of the leptons and quarks show a remarkable hierarchy. The third family quarks, t and b, together with the tauon, are the heaviest objects. The huge mass of the t quark dominates the spectrum.

Recall that at the beginning of the 20th century, an energy spectrum was observed in the hydrogen atom. It had a simple structure, but this structure could not be theoretically explained. The advent of quantum mechanics removed the shroud of secrecy. We are now looking for something similar to happen for leptons and quarks.

In the theoretical picture of the Standard Model, leptons and quarks are points, singularities in space. Can such infinitely small points even have a mass? If yes, then the question arises: why is the muon 207 times heavier than the electron? Both particles are identical, except for the mass. It could very well be that in the future we will be forced to give up the idea that these particles are point-like. The masses of the leptons and quarks could be a result of a substructure, just as the proton mass follows from its substructure. Possibly, future experiments with

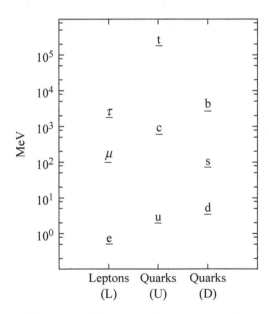

The masses of the charged leptons and quarks.

accelerators will produce evidence for a substructure. The leptons and quarks might consist of smaller building blocks.

But a glance at the clock tells me that we must adjourn our discussion. I have a meeting with a physicist here at Caltech, but I hope to be back in about 45 minutes. I recommend that you go for a walk around the campus.

Haller went to his meeting, and Einstein and Newton set off in the direction of the Beckman auditorium. Einstein spoke of his meetings in 1929 with Millikan at the Athenaeum. Millikan had wanted to convince Einstein to come to Caltech, and Einstein had shown interest. But then Einstein gave a speech on the radio about a trip he had made to the Soviet Union, in which the Soviet system came off looking quite positive. After that, Caltech's trustees decided to dump the offer to Einstein. Einstein decided to go to Princeton.

Einstein and Newton walked down the street towards the Millikan Library and then back to the Athenaeum. Einstein wondered about an old cannon standing in the middle of the campus. A student informed him that the cannon had been stolen by Caltech students years ago from the campus of the University of Los Angeles, UCLA. The cannon points directly towards the office of the president of Caltech. The president knows about this, but he also knows that the cannon is not loaded.

THE STANDARD MODEL AND THE NATURAL CONSTANTS

8

It was not until late that afternoon that Einstein, Haller, and Newton met again in the library of the Athenaeum to continue their discussion.

Haller: Thus far it has not been possible to bring together the interactions of the Standard Model with the gravitational interaction in a unified theory. The main reason for this is that the gravitational force, as Einstein formulated it, is not a physical force like the electric force, but an expression of the curvature of space and time.

Einstein: Correct — gravity is not a real force, at least not in my theory, but a manifestation of the curvature of space and time. This immediately leads to problems. In the case of the electromagnetic interaction, the move from the classical theory to the quantum theory can be made easily; in the case of gravity, it is an unsolved problem.

Haller: Yes, a quantum theory of gravitation is not in sight. It's obvious why. A quantum theory would mean that space and time are quantized, i.e., they obtain quantum properties. But space and time are continuous notions in our way of thinking. The Standard Model is formulated using a continuous time and space. If we give up this continuity and introduce a quantized space and a quantized time, I don't know what happens. There are a number of theoreticians who are working on this problem, though I never did.

It is possible to estimate that at very small distances, around 10^{-33} cm, Heisenberg's uncertainty principle destroys the usual space–time structure. If the electron is indeed a point-like object, then at this very tiny distance it would be "softened up" i.e., it would lose its point-like nature. Some theoreticians think that leptons and quarks are manifestations of small one-dimensional objects called superstrings. A small thread-like structure is indeed less singular than a point-like object, and as it turns out, there are fewer problems with a unified theory of quantum gravity when superstrings are introduced. However we have to accept that instead of the usual four dimensions of our space–time (three for space and one for time), we now have ten dimensions.

Einstein: Ten dimensions? That sounds crazy.

Haller: On the surface, this seems in conflict with our daily experience. But it can be arranged that of these ten dimensions, only four are macroscopically relevant. The other six come into play only at very small distances, when the effects of gravity begin to appear. A simple model will illustrate this. If we roll a piece of paper up tightly, it will look, from a large distance, like a one-dimensional object. The second dimension is not seen until we examine distances that are comparable to the thickness of the paper. Analogously, we can imagine the six dimensions to be rolled up. Should our space have extra dimensions on a very small scale, then this could help us to understand certain phenomena in particle physics.

To date it is not clear why, but nature seems to have a preference for the number three. There are, aside from the three dimensions of space, three families of leptons and quarks, and the quarks themselves have three colors. So the number three itself appears three times, and we have not the slightest idea why. The additional dimensions might explain some of these facts so that, in the end, some of the phenomena associated with the fundamental particles could be explained geometrically. You explained gravity in a geometric way, Mr. Einstein. Perhaps parts of the Standard Model can also be explained in a geometric way.

Einstein: Well, if these new dimensions manifest themselves in that way, I don't mind. My friend Hermann Weyl once wrote a book entitled *Space, Time, Matter*. It seems that he wished to express something similar. In his book he introduced one additional dimension and used it to generate the electromagnetic interaction besides gravity. But I guess that he wasn't thinking about six extra dimensions.

Haller: If and how such a geometric way of looking at the basic concepts of particle physics will ever be successful is still a mystery. Perhaps the idea of the superstrings makes sense.

Einstein: This seems pretty peculiar to me. To be honest, I don't believe in superstrings. Nature must somehow be simple, simpler than this theory of little worms called superstrings.

Newton: Mr. Einstein, I can't agree with you here. The theory of superstrings might be right. Who knows? It is at least a possibility. People nowadays aren't that dumb.

I must say, Mr. Haller, that the rise of the Standard Model is really exciting for me. I know from reading some books how particle physics looked shortly after the Second World War — it bears no comparison to today. After all, the Standard Model allows us to explain new phenomena in nature; it is quite an advance since my *Principia*. In a new edition of the *Principia*, I would like to include the Standard Model.

Haller: Without doubt it is a great theory. We have found simple principles that govern the forces of nature. They are described by gauge theories. This does not mean, however, that today's model represents an end to research in basic physics. Important questions are not answered in this model. In addition to the unsolved problem of understanding the particle masses, there is the question of the origin of the fundamental constants in the universe.

Einstein: Exactly. You said earlier that a large number of these constants are needed. That immediately makes the theory less attractive. How many constants are there in the Standard Model?

Haller: There are 28 constants, and we shall discuss them soon. Before we do that, I would like to say something about space and time. As a child, I thought that the Newtonian law of gravitation was easy to understand. The decrease of the force by $1/R^2$ corresponded to the increase in the surface area of a sphere by R^2. I later heard that it was Immanuel Kant who realized this. He was not only a great philosopher but also a good scientist and a great admirer of yours, Mr. Newton.

Kant realized that if space had four dimensions, the Newtonian law would imply a decrease like $1/R^3$. So Kant was the first to see the connection between the number of dimensions and the gravitational force — a remarkable insight for a philosopher.

Einstein: Indeed, that's not bad for a philosopher. Most philosophers don't even know that space has three dimensions. Kant should have become a physicist.

But now to the dimensions of our space. Our three-dimensional space has a special property. Even Plato noticed how remarkable the difference between two and three dimensions is.

In two dimensions there is an infinity of regular polygons, i.e., the triangle, square, pentagon, etc. This is not so in three dimensions. There are only five regular three-dimensional polyhedrons. First we have the tetrahedron, with three triangles as sides and four corners; then the cube, with six square sides and eight corners; the dodecahedron, with 12 five-

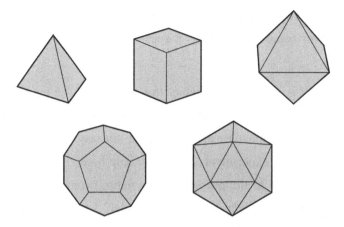

The five Platonic solids.

sided sides and 20 corners; and finally the icosahedron with 20 triangular sides and 12 corners. There are no others.

Also, as my friend Paul Ehrenfest noted around 1917, a stable orbit is only possible in three dimensions. He also discovered that in a world of more than three dimensions, there would be no stable atoms.

Haller: That's right, but that's only true if the new dimensions are really there on a macroscopic scale. In the new theories, we can't see them because they're rolled up. These new dimensions are often thought of in relationship with the degrees of freedom that we have in physics, the number of quarks and leptons, for instance.

Einstein: How does the theory work in detail? Is it possible to calculate anything, such as the masses of the quarks and leptons?

Haller: No. No one even knows if these theories make sense. But it would be interesting to have, finally, a purely geometric view of the world. Then everything would be contained in the space or the space–time. That would be something for you, since you have always loved the geometric aspects of our world so much.

Einstein: I have nothing against this proposal, but I do think that it should deliver some new insight. I guess a lot of work has still to be done.

Haller: Yes. It is hard to say whether this approach is on the right track. It might not be. Doing research can sometimes mean going down the wrong path, but we must ensure that we do not stay on the wrong path too long. Only new experiments can really help here.

But now we turn to the problem of the natural constants. How many constants are there in today's Standard Model? I will take for granted that we have only three dimensions of space and one time dimension.

Einstein: One moment. There are a number of particles in the Standard Model that we don't really need for our daily lives, such as the muon and b quark. I don't need a muon to eat and sleep well. I think Rabi was right to ask, "The muon, who ordered it?"

Let us forget about these particles and count only those that we really need: the u and d quarks, electrons, and all the force carriers (photons, gluons, the W and Z particles). How many constants does that leave? It can't be that many.

Haller: We can work that out quickly. First I would like to add a constant that Newton introduced, the constant of gravitation. It does not appear in the Standard Model, but we should not forget it. In this case, we have the constant of gravitation, the fine structure constant for the electromagnetic interaction, the coupling constant for the strong interactions, the mass of the electron, and the masses of the two quarks — that makes six constants to describe the stable matter. That's not so bad.

But that is not the end yet. We still have the weak interactions. Here we have a constant analogous to the fine structure constant, the mixing angle that we've already mentioned, and finally the mass of the W boson. Altogether, we have nine constants.

Newton: That is already quite a lot. I would expect a fundamental theory to have far fewer fundamental constants. After all, these nine parameters all have to be determined by experiment.

Haller: Indeed, they have to be determined by experiment. But let me say something about the Higgs mechanism, which we mentioned earlier, to bring the mass parameters into the focus of the theory.

The Higgs mechanism was originally introduced to generate the mass of the W particle. With the Higgs mechanism, it becomes possible to introduce the mass without disturbing the renormalization, as realized by 't Hooft and Veltman. This is because the mass is now a result of a broken symmetry, as in the example with the Mexican hat. If the symmetry is exact, the W particles have no mass. If the symmetry is broken, the mass is introduced, but by the back door, so to speak, and renormalization is preserved.

Einstein: I am not convinced. What about the mass of the electron? How is it generated?

Haller: This is very easy. The hypothetical Higgs field undergoes an interaction with the electron, and the mass is produced through this interaction.

Newton: Does that mean you can actually calculate the electron mass?

Haller: No. All we know now is the strength of the interaction between the Higgs particle and the electron. The electron mass remains a free parameter.

Einstein: That is not convincing. One free parameter is just replaced by another, and you end up with a theory that says absolutely nothing about the mass of the electron. I suspect the theory is not right.

Haller: I think along the same lines. We don't know if the Higgs mechanism is correct, but hopefully it will be possible to say something more about it in the near future, with the help of results from the new

LHC accelerator. The experimentalists are hoping to find the Higgs particle. Well, we will see what happens. Personally, I have my doubts about their chances. I believe the mechanism of mass creation is different from the Higgs mechanism. How things are in detail, I don't know, and I don't know anyone who does.

Let us now count the fundamental constants. There are six constants for the six quark masses, and another six constants for the six lepton masses. The neutrino oscillations are described by three mixing angles. In addition, one needs three complex phase parameters. Thus for the leptons we need 12 constants.

The quarks appear in the weak interactions as mixed states. For example, a c quark does not, as one would naively expect, turn with 100% probability into an s quark when it decays by emitting a virtual W boson. It does this only with 95% probability; there is a 5% chance that it will turn into a d quark. Thus the quark emitted in the weak decay of a c quark is not a mass eigenstate but a mixture of d and s. This mixing forms a bridge between the first and the second family. For the six quarks we need three rotation angles.

Newton: This is just like the rotation of a vector in three-dimensional space, which can be described by the three Euler angles.

Haller: Yes, however in the case of the quarks we are dealing with complex fields. They are described by complex numbers. Thus we have, besides the three mixing angles, a complex parameter. This complex parameter is important for the physics because it describes the violation of the CP symmetry.

Einstein: For heaven's sake, what's that?

Haller: It is a simple symmetry connected with antiparticles. When we go from particles to antiparticles, we make a "C-transformation". We can also perform a parity transformation at the same time, called a "P-transformation". This means that we exchange left and right. If we make the two transformations at the same time, the result is a CP-transformation. Even after the discovery of the violation of parity it was assumed that the CP symmetry would be perfectly preserved.

It was found, however, that CP symmetry is also violated, albeit a lot less than C or P symmetry is violated. In the Standard Model this violation can be simply described by a complex parameter. So far, the experimental results are well described by this simple theory.

Einstein: It still sounds complicated. Why is the CP symmetry violated at all?

Haller: We have known since 1964 that the CP symmetry is violated. The violation was observed in Chicago. At the time, nobody believed that such a violation would be found, but there it was.

If we include the complex parameter, we have six quark masses, three mixing angles, and the aforementioned parameter. Thus we obtain ten parameters. The same number of parameters enter for the leptons. But here one expects two further parameters that describe solely the neutrino masses. Thus the quarks and leptons are described by 22 parameters.

We also have the three parameters for the interactions — the strong, electromagnetic, and weak interactions — plus the constant of gravitation. Furthermore, the mass of the W boson and the mass of the hypothetical Higgs boson enter as free parameters. Altogether, we obtain 28 fundamental constants.

Einstein: I am slowly getting a headache. If there are 28 parameters, that means there are 28 numbers that the experimental physicists have to determine. This does not look to me like a great theory.

Newton: Perhaps the Standard Model is just part of a larger theory. It could well be that there are further interactions and that some of the seemingly free parameters may turn out to be calculable.

Haller: Perhaps, nobody knows. Now it's time for dinner. I suggest we adjourn. Let's meet in the restaurant.

Einstein: That's a good idea. I'm getting hungry hearing about all these mysterious natural constants, a nice steak would be great.

Einstein, Haller, and Newton soon met in the restaurant of the Athenaeum.

THE GRAND UNIFICATION

9

The next morning the three physicists had a big American breakfast in the Athenaeum. Afterwards they reconvened in the library.

Haller: I would like to discuss now the idea of embedding the Standard Model in a larger theory. This was suggested by theorists around 1975, and it allows us to reduce the number of fundamental constants.

Einstein: What generalization of the Standard Model do you have in mind? The Standard Model is all right, even if it does seem to me a little obscure.

Haller: The Grand Unification takes us away from the physics of the observed elementary particles. We enter a purely theoretical area. Nobody knows if this approach makes any sense. We may be on the wrong track. What we are trying to do is to embed the electroweak and the strong interactions in a unified theory.

Let me first point out a problem of the Standard Model. It is obvious that there is a profound connection between the charge of the electron and the charges of the quarks. The charge of the d quark, for example, is exactly equal to the electron charge divided by three. But this connection is not present in the Standard Model, as we have already remarked. There is nothing in the model to stop us from arranging other charges for the quarks.

In the electroweak interaction we have a symmetry, described by the group SU(2), called the weak isospin. The leptons and quarks appear as doublets of the weak isospin. For the strong interaction we have the color symmetry of the quarks, described by the group SU(3). If we could succeed in unifying the two symmetries, then we could probably fix the charges of the leptons and quarks. The first attempt in this direction was made in 1974 by Howard Georgi and Sheldon Glashow at Harvard University. But their model, based on the symmetry group SU(5), is now excluded by experimental data, as we shall see later.

In the summer of 1975, Harald Fritzsch and Peter Minkowski, and, independently, Georgi found that the group SO(10) would be a good candidate for the unification of the basic interactions. That summer Fritzsch and Minkowski were visiting the Aspen Physics Center in Aspen, Colorado. This is a beautiful institute in the midst of the Rocky Mountains. Physicists often visit the institute during the summer months with their families, combining vacation and work.

Fritzsch and Minkowski unified the leptons and quarks by considering the leptons as a fourth color. This is, of course, only possible formally. The corresponding symmetry must in reality be strongly broken, or there would be no leptons in nature.

Adding the leptons as a new color, we end up with four colors, i.e., the color group is now SU(4), the symmetry group of a complex four-dimensional space. A group, by the way, is a mathematical concept. It is a set of symbols for which the product of two symbols is defined in an abstract way. A simple case is the group of whole numbers: the sum of two whole numbers is a third whole number.

Einstein: I know what a group is, but Newton doesn't. Let us take the group SO(3) as an example. SO(3) is the group of rotations in three-dimensional space described by the three Euler angles. The product of two rotations is also a rotation. The group SU(3), which we have discussed in more detail, is a group of rotations in a three-dimensional complex space. Each group element is described by eight real parameters. The group SU(4) describes rotations in a four-dimensional complex space.

This reminds me of something peculiar that my friend Hermann Weyl taught me about the group SU(4). I learned that it is isomorphic to the group SO(6) — in other words, it is practically the same as SO(6). This means that SU(4) can be viewed as the symmetry of rotations in a six-dimensional space described by real numbers. But does this have anything to do with the group SO(10)?

Haller: I'll get to that, but first I want to return to the weak interaction. The group that plays a role in weak interactions is the group SU(2), the group of rotations in a two-dimensional complex space. When unifying the weak and electromagnetic interactions, it turns out to be useful to consider the larger group SU(2) × SU(2), one SU(2) for the right-handed fermions and one for the left-handed ones.

Einstein: Mr. Newton, this group is what you get by performing two rotations in two dimensions. We rotate the right-handed fermions and the left-handed ones independently of each other. This makes me think again of Hermann Weyl, who explained to me that the group SU(2) × SU(2) is isomorphic to the group SO(4), the rotations in a four-dimensional space.

Haller: Mr. Einstein, you now have all you need at your disposal. This is exactly how Fritzsch and Minkowski were thinking in Aspen. On the one hand, they had the group SO(6), on the other hand the group SO(4).

Thus the product SO(4) × SO(6) appeared, and suddenly the group SO(10) was in the air. It turned out that leptons and quarks could indeed be described by the group SO(10).

Let us consider only the u and d quarks, the electron and its neutrino, and the corresponding antiquarks — the particles of the first family. These are 16 objects when we pay attention to the colors of the quarks and antiquarks: three u quarks, three anti-u quarks, three d quarks, three anti-d quarks, the electron, the positron, the neutrino, and the antineutrino. Let me write them in a large circle.

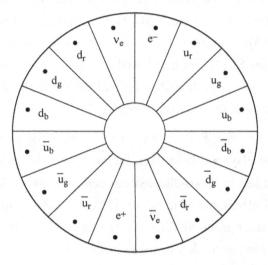

Particles in a 16-dimensional representation, described by the group SO(10).

Now let us imagine that there exists an interaction capable of changing all 16 objects into one another. The interactions of the Standard Model are something like this. A W particle, for example, changes a u quark into a d quark. The group SO(10) describes exactly how these interactions happen.

Einstein: I have a problem with that. The 16 particles that you have just mentioned include electrons, positrons, and quarks. Thus there must be an interaction that changes quarks into electrons or positrons. It would mean that a proton, which consists of three quarks, could decay into a positron and a photon. Is this possible?

Haller: Indeed, you are right. The only way to suppress these interactions is for the masses of the particles that mediate them to be large. It is easy to calculate a lower limit for the masses, and we find that it must be around 10^{16} GeV or higher.

Einstein: Oh God, that is gigantic. That's almost equivalent to the mass of a bacterium. Does that make sense?

Haller: When I heard about this for the first time, I was skeptical. But I have since gotten used to the large energy. We can arrange the SO(10) theory in such a way that all the interactions from the Standard Model are there and all the new interactions, beyond those found in the Standard Model, are very weak because of the large masses of the particles mediating the new interactions. So there is no conflict between the model and experiments to date.

Einstein: But these new interactions are there. Does this mean that experimentalists should look for decaying protons?

Haller: Yes! Physicists started building big detectors to look for proton decays after about 1975. An interesting decay that happens in the SO(10) theory, for example, is the spontaneous decay of a proton into a neutral pion and a positron.

The largest detector for proton decay was built in a mine in Japan in a place called Kamioka, a small village south of Toyama. The name of the detector, Superkamiokande, is derived from the name of the village. I was there in 1997 and visited the detector.

My friend Yoji Totsuka, the director of the institute, drove me through a long tunnel to the detector in the middle of the mountain. The detector consists of a large pool of clean water. The water is surrounded by tens of thousands of photomultipliers. Particles emitted in the decay

The Superkamiokande detector near Kamioka, Japan. In the photograph one sees a boat with people cleaning photomultipliers.

of a proton would produce a flash of light, which could then be observed by the photomultipliers.

Einstein: Have they observed a particle decay?

Haller: Not so far. They can now give a rather large limit for the lifetime of a proton. It is about 10^{32} years. But they keep looking.

Now back to the theory. Let us first look at the group SU(5) of Glashow and Georgi, which describes transformations in a five-dimensional complex space. This is the smallest possible group that can contain the Standard Model.

The fermions of the first lepton–quark family are:

$$\begin{pmatrix} v_e \\ e^- \end{pmatrix}(e^+)\begin{pmatrix} uuu \\ ddd \end{pmatrix}(\overline{uuu})(\overline{ddd})$$

Now we divide the fermions into two systems. Five fermions are placed in a 5-representation:

$$\begin{pmatrix} v_e \vdots \overline{ddd} \\ e^- \vdots \end{pmatrix}$$

The other ten fermions are placed in a 10-representation of SU(5):

$$\begin{pmatrix} uuu \vdots \\ ddd \vdots \end{pmatrix} \overline{uuu}, e^+$$

The fermions of the other two families can be described analogously. As you see, there are 15 fermions in one family. We have 15 fermions instead of the 16 mentioned before in connection with SO(10) because in SU(5) the neutrino only appears as a left-handed particle.

Einstein: I prefer the SO(10) theory, which has only one representation.

Haller: I also prefer the group SO(10), but we are just examining the SU(5) theory as an example. The unification can tell us something about the electric charges. The electric charge must be one of the charges of the SU(5) group. This group has $5^2 - 1 = 24$ charges. Note that the group U(1) has only one charge, the group SU(2) has three, and the group of the color charges SU(3) has eight charges. The group SO(10) has 45 charges.

The charges of a group have the interesting property that the sum of the charges of the elements of a representation must be zero. So the charges of the five fermions in the 5-representation that I just described must add up to zero. Since the charge of the neutrino is zero, we obtain a relationship between the electric charge of the electron and the electric charge of the d quark:

$$Q(e^-) = 3Q(d).$$

Einstein: So we find exactly the electric charges observed in nature. The factor three is the number of quark colors. I thought that the triality of the charge and the color of the quarks might be closely connected, and here it comes out that way. If there were four colors, you would have to replace the number three by four. The charge of the d quark would then be $-\frac{1}{4}$.

Haller: Yes. By the way, in the SO(10) theory it is just the same. In a similar way, we find the charge of the u quark to be $+\frac{2}{3}$. Its charge is exactly the charge of the d quark plus one.

The SU(5) theory makes a concrete prediction about the decay of the proton. It predicts that the decay should take place with a lifetime of about 10^{32} years.

In the framework of the SU(5) theory, we can also calculate the electroweak mixing angle Θ_w and the chromodynamic fine structure constant. We find:

$$\Theta_w = 37.8°, \ \alpha_s = \tfrac{8}{3}\alpha \approx \tfrac{1}{51}.$$

Einstein: Good, so now we have fewer free constants. Are these values correct?

Haller: Here there is a problem. The predicted values are not the same as the values found in experiments. The angle Θ_w, for example, is measured to be around 28.7°, not 37.8°, and α_s is not as small as predicted.

There is a further problem with the SU(5) theory. This theory has 24 gauge bosons, corresponding to its 24 charges. These include the eight gluons of QCD, the two W bosons, the Z boson, and the photon. The other 12 gauge bosons are new bosons, and they are associated with new interactions. These interactions can turn a lepton into a quark, for example, as discussed before, and protons can decay.

In the SU(5) theory the lifetime of the proton depends on the masses of the 12 new gauge bosons. The observed stability of the proton therefore tells us that the masses of these particles must be enormous, at least 10^{16} GeV. That is the only way the proton can live more than 10^{31} years, as observed. It should be noted that this age is much greater than the age of the universe, which is about 14 billion years.

If the SU(5) theory is correct, the coupling strengths of the strong, electromagnetic, and the weak interactions at energies above 10^{16} GeV will all be the same. This is because the different interactions are simply different manifestations of one unified theory.

For this reason, we expect that around energies of 10^{16} GeV the mixing angle will be 38° and the relationship $\alpha_s = \frac{8}{3}\alpha$ will apply between the strong and electromagnetic interactions.

But I have to stress that our knowledge of the interaction coupling strengths is derived from experiments at relatively low energies, $E \leq 10^2$ GeV. We cannot expect the coupling strengths at these energies to be the same as those expected around 10^{16} GeV. In the framework

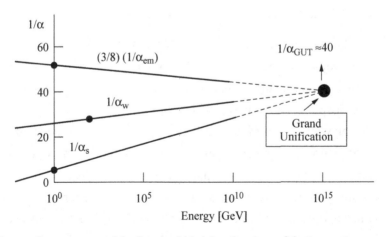

The coupling constants of the Standard Model as functions of the interaction energy.

of quantum theory, we expect the coupling strengths to vary slowly as a function of the corresponding energy scale. That is why the coupling strength in QCD becomes smaller as the energy increases (due to asymptotic freedom), while in a U(1) gauge theory the coupling strength gradually increases.

With the help of the LEP ring at CERN, it has been possible to determine very precisely the parameters of the electroweak theory at low energies. We can extrapolate this knowledge of the coupling strengths to higher energies. We find that the three coupling strengths do come close to each other around 10^{16} GeV, but they do not meet in a single point as demanded by the SU(5) theory.

The behavior of the coupling constants as a function of the energy scale definitely indicates that the idea of a unification of the fundamental forces is sensible. Only it seems that on the long road from the energies accessible today to the energies where unification is predicted to take place, something more than a slow change of the coupling parameters as a result of their interactions must occur. It could be, for example, that at around 1000 GeV new symmetries and new interactions appear. A number of theoreticians nowadays are discussing the appearance of a supersymmetry.

Einstein: Oh God, what kind of symmetry is that?

Haller: The symmetries we have dealt with so far, isospin, for example, contain multiplets of particles that all have the same spin. A symmetry transformation can change a proton into a neutron, but it cannot turn a proton with spin $\frac{1}{2}$ into a meson with spin 0.

In supersymmetry, however, it is possible to turn particles with spin $\frac{1}{2}$ into particles with integer spin. Such a transformation could turn a quark into a particle with zero spin. Such a particle has not been seen in experiments, but there could be a new particle with a mass too large for us to have been able to observe it so far. This hypothetical particle has been dubbed a "squark". It's an awful name, actually, but it refers to the supersymmetric partner of a quark. Indeed, in the supersymmetric variant of today's Standard Model, there is a "new" boson for every "old" fermion and a "new" fermion for every "old" boson. The supersymmetric partner of the photon, which has spin $\frac{1}{2}$, is the hypothetical photino.

Today it is still not clear whether supersymmetry is really present in nature. If it is, there must be a critical energy level above which super-symmetry comes into play. This critical energy would set the mass scale for the supersymmetric particles. Today we expect that supersymmetry, if it exists at all, starts at an energy of around 1000 GeV.

If supersymmetry is present in nature, the supersymmetric particles will appear in the interactions, thereby changing the behavior of the coupling parameters at very high energies. It turns out that, with the presence of supersymmetric partners above 1000 GeV, the coupling strengths do converge to a point at about 2×10^{16} GeV. So a supersymmetric version of the SU(5) theory is consistent with today's experimental values for the coupling parameters. In the supersymmetric version of the SU(5) theory the proton is still unstable, but it lives a little longer than in the non-supersymmetric version of the theory, around 10^{33} years. The experimental limits are compatible with such a long lifetime.

Time will tell if nature really makes use of supersymmetry. I am skeptical; I don't believe in it. With the SO(10) theory, a unification of the interactions is possible and consistent with the experimental data without the introduction of supersymmetry.

Einstein: I actually have nothing against decaying protons. They can be quite useful in cosmology, right?

Haller: Yes, if the proton really can decay into leptons and photons, it means that the baryon number is not exactly preserved. That would help to explain one of the strange phenomena in our universe.

The matter in our world is composed of electrons and nucleons, and the latter are composed of quarks. Antimatter made up of antiquarks doesn't seem to be present in our world. The stars in our galaxy are made up of matter, not of antimatter. There is also evidence that distant galaxies are made up of matter, not of antimatter. Thus the baryon number of the visible universe is huge.

We know that the universe was created about 14 billion years ago in the Big Bang. If the baryon number were strictly preserved, it would have been as large as it is now even shortly after the Big Bang. This would mean that our cosmos had been born with a very large baryon number. That doesn't make much sense. It would seem a lot better if the baryon number were zero at the beginning. Thus the number of quarks and antiquarks would have been equal at the start. This is possible if the baryon number is not strictly conserved, as is the case in the SU(5) theory and the SO(10) theory.

The new interactions that change the baryon number would allow the baryon number to be zero at the Big Bang but huge today, making the baryon number a product of history. One possibility is that right after the Big Bang there were about 10 billion antiquarks per unit volume and 10 billion-plus-one quarks. Thus there was only a very small violation of the CP symmetry. But soon after, the quarks and antiquarks annihilated, and only the quarks which did not find an antiquark to annihilate stayed in the universe. These would be the quarks in our world.

Einstein: This sounds convincing. Now I have a question. You spoke earlier of the unification of the interactions. You have to unify the strong interaction and the electromagnetic interaction, but their strengths are completely different. How is that supposed to work?

Haller: That is a very good question, and the answer will turn out to your liking. But let us wait a while. This morning I obtained a key to the garden at San Marino from the secretary at Caltech. It is just around the corner, and we could take a walk in the park now. I suggest that we leave right away.

Thus the three physicists left the Athenaeum and went to the end of Hill Street. Then they walked down Arden Street towards San Marino. They reached a small crossroads and found themselves at the entrance to the big garden of San Marino.

In the Garden of San Marino

10

It was a beautiful afternoon. Haller opened the small gate with his key, entering into the Huntington garden at San Marino. The Huntington garden is a botanical garden named after Henry Huntington, who began planting it in 1903. It is a very beautiful park with many exotic trees and bushes. Amongst other things it contains a Chinese garden, the largest one outside of China, a Japanese garden, a desert garden, a palm tree garden, a rose garden, and a jungle garden. All the different types of vegetation of the American continent are represented. In the eastern part of the garden one finds a museum and a library.

They strolled through the garden until they were near the museum, then they sat down underneath a large pine tree.

Einstein: This is a nice place. It's actually far too beautiful here to be discussing something as mundane as physics. But I suggest that we stay here for now and that Mr. Haller finally answers my question about the strengths of the interactions.

Haller: That I will gladly do. It follows from quantum theory that the constants of the interactions are not constant, but depend on the energy at which the interaction occurs.

Let us look at the fine structure constant. It is normally measured at very low energies, basically at zero energy, but it has also been measured at LEP at high energy, at about 200 GeV. This is almost 400,000 times the energy associated with the mass of the electron. The strength of the electromagnetic interaction measured at this high energy was larger than that at low energy: $\frac{1}{127}$ instead of $\frac{1}{137}$. This is exactly what is predicted by the theory of quantum electrodynamics.

Newton: It's funny that the strengths of the interactions are not really constants. So my constant of gravity will not be a real constant either. In the example you've given, the fine structure constant changes by about 10%, which is quite a lot. Why are the strengths of the interactions not constant?

Haller: In quantum theory, every particle is surrounded by particle–antiparticle pairs. An electron, for example, is surrounded by electron–positron pairs — the electrons are pushed away, the positrons are attracted, and thus the charge is partially shielded. The shielding becomes less effective at high energies because we reach closer to the core of the electron and so do not see some of the pairs. Thus the interaction becomes larger.

The change of the interaction constant in QCD is interesting. In this theory, the strength of the interactions decreases at high energies. Gluons do not only interact with quarks, they also interact with other gluons, and these interactions weaken the strong force at high energy, as we have already mentioned. This is the effect of asymptotic freedom. Thus the constants describing the strength of an interaction depend on the energy or on the scale.

Einstein: Let us take another look at the natural constants. About 20 of them have to do with the masses of the leptons and quarks. Then there are the constants for the three interactions, and some constants related to the Higgs mechanism. The masses of the leptons and quarks are the constants that I wonder about most. I find it very strange that nobody can calculate the electron mass.

Haller: Yes, I agree. It is strange that no theoretician can say anything about the mass of the electron. By the way, it's not the absolute mass that is important, but its ratio to other masses, such as the ratio of the muon mass to the electron mass, or the ratio of the proton mass to the electron mass.

Einstein: A while ago you said that leptons and quarks might be composed of even smaller units. Might that help? After all, nowadays the proton mass can be calculated from the properties of the quarks it contains, using QCD. If smaller units exist inside leptons and quarks, perhaps we could then calculate the masses of the leptons and quarks.

Haller: Today speculating about further substructure is not popular. But in my opinion we should continue searching in this direction, even if the idea of substructure is difficult to combine with the idea of Grand Unification. We need an experimental signal.

We have looked for substructure in accelerator experiments, for example with HERA at DESY in Hamburg, but so far without success. The new LHC at CERN, however, might soon find something. I have looked into models of substructure, but I have found no satisfactory way to calculate the masses.

I would like to address two other problems now, both of which are concerned with the creation of our universe by the Big Bang. In the beginning, the universe was very hot — all particles had a very high energy — and both quarks and antiquarks were present.

Einstein: We expect that shortly after the Big Bang matter and antimatter were present in equal proportions. Shouldn't there be a lot of antimatter in the universe today?

Haller: Today there seems to be no antimatter in the universe, or at least none that we can see. There cannot be any antimatter stars in our galaxy, because otherwise we would see violent fireworks when normal stars and stars consisting of antimatter collide. Distant galaxies cannot be made of antimatter either, because we often see galaxies collide without any fireworks.

It is easy to see that a universe that was totally symmetric with regard to matter and antimatter would not really make any sense. A large part of the matter and antimatter would have annihilated, and the remaining mass should be as much antimatter as matter, which is not what we observe.

We do occasionally see a few antiprotons in cosmic rays, but they were probably produced in high energy collisions of protons or nuclei. We have never seen a complicated atomic antinucleus in cosmic rays. The only way out of the trap of antimatter is a breaking of the symmetry between matter and antimatter, i.e., a breaking of the CP symmetry.

Newton: This symmetry is broken anyway, experiments have shown that. Is the observed violation in accordance with cosmology?

Haller: This we don't know for sure, but it seems to be in agreement. The fact that CP symmetry is violated seems to fit with the observation that there is matter but no antimatter in the universe.

But now let us move to another problem. Shortly after the Big Bang there were primarily two elements in the universe, hydrogen and helium. Hydrogen is made up of one proton and one electron. It is clear that hydrogen will form because protons and electrons are present in the universe soon after the Big Bang. Helium forms through a simple nuclear process. Its nucleus is an alpha particle, made from two neutrons and two protons. We can easily calculate the details of this process, and the results fit well with what is observed. We find that helium, on average, accounts for 24% of the nuclear matter in the universe.

Heavier elements, such as carbon, can only form through the combination of hydrogen and helium nuclei. But now it gets complicated. We need three helium nuclei, i.e., alpha particles, to make a carbon nucleus, but collisions of those three particles are very rare. This process is called the triple-alpha process.

Einstein: Yes, I recall. In 1952, the English astrophysicist Fred Hoyle brought our attention to this process, in particular to a serious problem. Were carbon to be formed by this mechanism, one would expect that most of the carbon nuclei would undergo another nuclear reaction with one of the alpha particles, creating oxygen. Hoyle studied the subject in detail and came to the conclusion that the carbon in the universe could only have formed if the process of carbon creation was very fast and efficient.

There is only one way to make a nuclear reaction more efficient: the formation of carbon nuclei could only be understood if there were a resonance in the vicinity of the energy given by the incoming nuclei. After some calculations, Hoyle came to the conclusion that the carbon nucleus must have a resonance at an energy of 7.65 MeV. Nothing was known from nuclear physics about such a resonance.

Haller: Nuclear physicists at Caltech played an important role in addressing this problem. New experiments done at Caltech by William Fowler and Charles Lauritsen showed that Hoyle was right. Fowler and Lauritsen found a prominent resonance at 7.656 MeV, which was exactly the energy predicted by Hoyle.

Newton: Still, the story seems a little odd to me. If nuclear physics were just a little different, then there would be no resonance effect, and then there would be no carbon and no life on Earth. Apparently our life depends on strange coincidences such as a nuclear resonance in carbon.

Haller: Mr. Newton, our life is linked to all kinds of strange coincidences. The energy of carbon's nuclear resonance depends on the natural constants. The strong interaction and the electromagnetic interaction play a role here. If the fine structure constant were just 4% different to what is observed, the position of the resonance would be different, so there would be no resonance effect and no life on earth.

Einstein: I believe the laws of nuclear physics and the values of the natural constants are predetermined. It does not make much sense to consider other values of the fundamental constants.

Haller: Mr. Einstein, perhaps that's how it is, but I doubt it. Well, we can't cover much more today, so let's finish our discussion.

Now, the plan for tomorrow. Haven't you always wanted to take a look at the big accelerator, SLAC, up at Stanford University? I suggest that we drive up there tomorrow morning, setting off after breakfast at around 8:30. It is a long drive to Stanford. Tomorrow we will drive to Lompoc, north of Santa Barbara.

Einstein: Fine, that's an offer we cannot refuse. A look at SLAC will be very interesting for us. I'll be happy to come.

They left the Huntington garden and walked back to the Athenaeum for dinner.

ON THE ROAD TO
EL CAPITAN BEACH

11

The next day the three physicists met in the Athenaeum's parking lot with their luggage. Haller arrived with a rental car, and they left the premises of Caltech. Einstein had insisted on driving. Haller asked Einstein for his driving license, and Einstein really had one! He had just received it a couple of days ago in Pasadena. Einstein drove north along Hill Street until they reached Interstate 210, which they followed west.

After more than an hour they reached the coast of the Pacific Ocean. Einstein insisted that they stop for a while and walk on the beach. They strolled along the beach for two hours.

Finally, they continued their drive in the direction of Santa Barbara. They passed the city after about an hour. Haller intended to stop at the El Capitan state park just north of Santa Barbara. In the 1970s, he had often camped there with his wife and his young son. They soon reached the state park. Haller paid the entrance fee, and they were at an empty campsite in a matter of minutes. Haller told his companions that he planned to take a long break here. Within a couple of minutes they had found a nice spot on the beach. They took a refreshing dip in the Pacific Ocean. Then Haller started the discussion.

Haller: Gentlemen, in spite of the nice weather and the fabulous surroundings, let us get back to physics. Today I would like to discuss an aspect of the fundamental constants that we have not yet considered. We have assumed that these constants are constant in time and throughout space. But are we really sure about this? It could be that the constants change slowly in the course of time.

Let me first discuss some results from the reactor at Oklo. An interesting discovery was made in June 1972 in a reprocessing plant for nuclear reactors in France. Uranium ore from Gabon, a republic in West Africa, was being examined. The ore was from a mine near the river Oklo, about 400 km away from the Atlantic Ocean. Normally uranium ore consists mainly of the isotope uranium-238, with an admixture of only 0.72% of the isotope uranium-235. An engineer making exact measurements of the ore from Oklo noticed that it had less of the isotope uranium-235 than expected. The question was, what had happened to the missing uranium-235?

After eliminating all other possibilities, the French physicists came to the conclusion that uranium-235 in the Oklo area had been destroyed by nuclear reactions. They concluded that there must have been a natural nuclear reactor at Oklo. The mining of uranium ore at Oklo was stopped for some time, and a geochemical survey of the area was done. This closer examination revealed that indeed there had been several reactors in the region two billion years ago. Fourteen areas in the vicinity of Oklo were identified where such reactors must have existed. More than two billion years ago, there were more atomic reactors in Africa than there are there now.

Einstein: As far as I know, the half-life of uranium-235 is 700 million years, while uranium-238 has a half-life of 4.5 billion years. That means that two billion years ago, there was a lot more uranium-235 than today.

Haller: Yes, when the Earth formed about 4.5 billion years ago, 17% of natural uranium was uranium-235. Two billion years ago, the ratio uranium-235:uranium-238 had gone down to around 3%. This is the value that is needed to start nuclear reactions. The process can be moderated by water, and at Oklo, water came from the river Oklo.

One interesting reaction is the capture of a neutron by a samarium-149 nucleus. This process creates a samarium nucleus with 150 nucleons and emits energy in the form of gamma radiation. The cross section measured for the reaction is very large. Nuclear physicists, in doing experiments with neutrons, discovered a resonance just above the threshold for this process that explains the large cross section. Without this resonance, the process would play hardly any role. The conclusion drawn from Oklo is that the position of this resonance cannot have changed during the last two billion years.

The resonance of samarium, like every nuclear resonance, depends on the value of fundamental constants such as the fine structure constant. If we assume that the fine structure constant changes, and nothing else, we conclude from the Oklo data that it could not have changed by more than 10^{-16} per year. This is quite an impressive result. But if we allow other constants to vary as well, it's a different story. It could be that both the strong and the electromagnetic interactions are changing with time but in such a way that their effects on the nuclear resonance cancel each other to a large extent. This may seem strange and improbable, but if we consider that the strong and electromagnetic interactions are somehow connected by a grand unification, it doesn't sound so crazy anymore.

Let me recapitulate. Examination of the natural reactors in Oklo has revealed that, if none of the other parameters change, the fine structure constant cannot have changed by more than 10^{-16}. If both the strong as well as the electromagnetic interactions have changed, the constraint on the time variation of the fine structure constant is not valid. This is expected if there is a grand unification of the interactions.

Einstein: Back to the question that is occupying my mind: what are the natural constants really? Where do they come from?

Haller: One of my American friends is of the opinion that they are cosmic accidents. He thinks that the constants fluctuated during the Big Bang and then, as the cosmos expanded, took on certain random values. If the Big Bang were to be repeated, there would be no chance of ending up with the same values for the fundamental constants. We would obtain a different universe.

Einstein: I recall that Paul Dirac thought about the time dependency of the constant of gravitation on his honeymoon, and that he wrote a paper on the subject. What happened to this idea?

Haller: It took a long time to test his idea. Very precise tests with satellites were done. Now we are sure that a time-dependency on the scale that Dirac had in mind — about an order of magnitude of change since the Big Bang, making a yearly change of about 10^{-10} — is not allowed.

However, physicists from Australia, England, and the United States have discovered in an astrophysics experiment, using the Keck telescope in Hawaii, that the fine structure constant has changed over time. However, other smaller experiments do not reproduce this result, thus things are not clear.

Newton: Still, another question comes to my mind. Shouldn't we be considering variations in space as well as in time? How do we know, for example, that the fine structure constant has the same value in some distant star?

Haller: We can measure atomic transitions in distant solar systems, in distant quasars, for example, or in the Andromeda cloud. These atomic transitions tell us how large the fine structure constant is at these locations. Within the limits of accuracy of our measurements, no variations in space have been observed. Even the experiment at the Keck telescope that found a variation in time found no variation in space.

Many different directions were explored. One observes a time variation but no space variation.

Haller: But let us leave now. It is time to get on the road again.

Einstein: Not so fast. It is so beautiful here at the El Capitan beach. I suggest that we take another stroll.

They left the state park on foot, going north past huge boulders. In some places they were forced to walk through water. After an hour they were back at the car. Einstein took the wheel again. Haller guided Einstein through the campsite. He could see the places where he had camped many years ago. Nothing had changed in the last 20 years. A few minutes later they were on the highway, and they soon reached the exit for Lompoc. Here they left the highway. Einstein drove down the country road until the first houses of Lompoc came into sight.

Haller knew a good Greek restaurant in Lompoc near a motel, and he gave Einstein directions to get them there. They first rented three rooms in the motel, then they went to the restaurant. They chose a corner table. The owner, obviously born in Greece, immediately offered them an aperitif which they gladly accepted. They ordered roasted lamb cutlets, the house speciality. The accompanying pita bread was delicious and Newton ate an incredible amount of it, at the same time drinking quite a lot of beer.

Einstein: My dear Newton, I can see that our trip to El Capitan has left you with quite an appetite and a parched throat. Enjoy your meal. Just don't drink too much beer, or we will have problems with the physics.

This is really an excellent place. I thought Lompoc would only have hamburgers and the rest of that awful stuff that Americans produce and eat.

Haller: I came through here for the first time five years ago on a drive from Stanford to Pasadena. I found this Greek restaurant. Since then, I have always made a stop here when traveling between Stanford and

Pasadena. Lompoc is pretty much in the middle. Usually I even stay a night here in a motel.

Einstein: During my time in California, there were practically no Greek restaurants, only Mexican places. I liked Mexican food though, such as a good enchilada or a burrito.

Haller: Then I have to tell you that we have really missed out on something. There is a very good Mexican restaurant in Pasadena called Acapulco. The enchiladas and the burritos at Acapulco are a pure joy. And the freshly baked tacos with good Mexican beer, you'd forget all your physics over those.

But for now we are here in a Greek restaurant, and the lamb cutlets, as you are about to discover, are really excellent.

They ate the delicious lamb cutlets and spoke mainly about American politics. Then Haller and Newton went to bed. Einstein, however, was not yet tired. He took a long walk through Lompoc. To his surprise, he nearly stumbled into a hyena. It came straight towards him down the middle of the street, stopping in front of him and staring at him with curiosity. After growling a little, the hyena went into the desert.

The night sky was very clear, as it most always is in California. Einstein looked for the Andromeda galaxy and found it quickly. He looked at it for a long time. The light now striking his eyes had taken two million years to reach him. Two million years ago, atoms in the stars of the Andromeda galaxy had radiated these photons. The photons had started traveling a long time before he was born and they were now reaching his eyes. What a crazy and wonderful world, he thought, full of coincidences. Although everything is determined by physical laws, our world is colorful and completely unpredictable, and that's exactly what makes it so beautiful. Einstein took a deep breath of the fresh Lompoc air and felt wonderfully at ease. Towards midnight he went to his room.

In Esalen

12

The next morning they got up early and had breakfast in a nearby café in Lompoc. Newton ordered an American breakfast with three eggs, lots of bacon, and French fries. Einstein and Haller just had a continental breakfast. Finally they were in the car traveling north, and soon they were alongside the Pacific on the famous Highway 1.

Einstein was going faster than the speed limit of 60 miles per hour, and they were soon stopped by a policeman posted in the middle of the woods. The policeman checked the driver's licence and was clearly amazed when he read the name "Einstein".

"Are you a relative of the great Einstein?" he asked Einstein. Einstein grinned and said, "I guess you could say that. How do you know of my grandfather?"

"How do I know of him? I studied physics at Stanford ten years ago, so I know him and his theory of relativity well. I stopped my study though," the policeman replied. He continued, "If you are a relative of the great Einstein, I will forget about your violation of the traffic laws. Here's your licence back, and please drive a little slower now. All the best."

They continued along Highway 1. After about three hours they reached the area of Big Sur, a very impressive region on the Pacific. Haller knew an institute, the Esalen Institute, beautifully located on the coast near Big Sur, where one could stay the night. The entrance to the institute was at a curve in the road. After waiting there briefly, they were allowed to enter. They could only get a three-bed room, but it had a beautiful view of the Pacific.

Haller had the idea of visiting a hot spring in the garden of the institute. They walked to a natural pool where they enjoyed bathing in the hot water.

Haller: If you don't mind, this afternoon I would like to approach the problem of the time dependence of α.

Einstein: That's fine with me. It's so comfortable here in the pool that I can even deal with a time dependent α.

Haller: I'll start with some observations that were made a few years ago, around the turn of the century. Today we are capable of studying the fine structure of atoms in distant galaxies and in distant quasars. A group of astrophysicists from Australia, England, and the United States made measurements of quasars with the Keck telescope in Hawaii. They examined the fine structure of atoms of iron, nickel, magnesium, zinc, silver, etc.

Newton: I didn't know that distant quasars contained silver. Do they perhaps even contain gold?

Haller: Quasars have all the elements that we have here on earth. But looking for gold doesn't make a lot of sense; it is just as rare in quasars as it is here.

Altogether, the team looked at 150 quasars, many of them billions of light years away, some even 11 billion light years away. A small time variation in α was found:

$$\Delta\alpha / \alpha = -(0.54 \pm 0.12) \times 10^{-5}.$$

The Keck telescope in Hawaii.

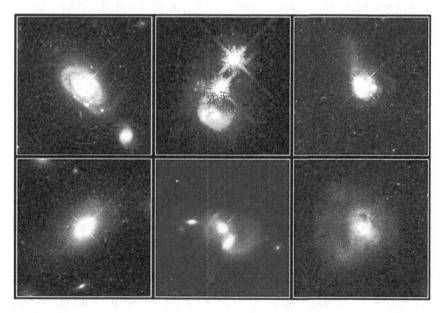

A quasar.

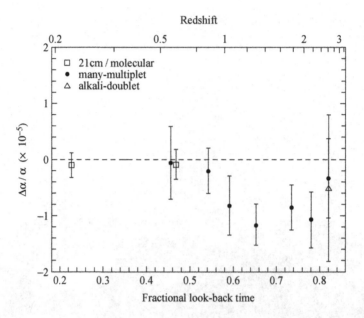

The variation of α with time.

Interestingly, the quasars were located in different directions in the sky, yet they all showed the same time variation. So we see a variation in time but no variation in space. If one assumes that the change is linear to a first approximation, this result suggests that the fractional change in α is about 1.2×10^{-15} per year.

Einstein: But it remains to be seen whether a linear approximation holds. Is it possible to look for a time variation in other experiments, e.g., in quantum optics experiments?

Haller: It is difficult, but it is not totally impossible. Laser physicists can make incredibly exact measurements nowadays. We shall discuss them soon.

Einstein: We should think about what it could mean if α were time dependent. We know that α is given by $e^2 2\pi/hc$, so it could mean that h, or c, or both are time dependent.

Haller: I'm amazed by what you are saying. Can the speed of light, c, be time dependent? What does that mean for your theory of relativity, Mr. Einstein? You destroy your theory.

Einstein: Ah, yes. In the heat of battle, I hadn't thought about that — I guess that it wouldn't be good for my theory, in fact, it would be terrible!

Haller: Yes, it would be fatal for your theory. I don't know what would be left of the theory of relativity if c were dependent on time. I just can't imagine it. I always tell my students in Bern to set c equal to 1. If c were time dependent, this could not be done.

Newton: The other day, I read in an astronomical paper that c might be time dependent, so there are scientists who take this seriously.

Haller: Don't waste your time on that. Astronomers don't have the faintest idea about the theory of relativity. They would be capable of performing relativistic calculations with a time dependent c. A time dependent c would be terrible for the unity of time and space on which the theory of relativity is based.

Einstein: Well, astronomers are star-gazers, not physicists, and they have no clue of the theory of relativity. They should not write any theoretical papers; these papers are always terrible. I agree that we had better forget the possibility of c being time dependent.

Newton: Well, all right. Can we make h time dependent, or does that lead to problems as well?

Haller: For me, that is also problematic. It would mean rewriting quantum theory. Then I would understand the theory even less than I do now! In particle physics, we often set $h/2\pi$ equal to 1, as well as c equal to 1, to simplify our calculations. That would no longer be possible.

Einstein: I agree with you. A time dependent h would make quantum theory even more mysterious. Let's forget a time dependent h. Then there is only one possibility left: the time dependency must be in e.

Newton: So e is time dependent? The electric charge is supposed to change in time?

Haller: That would be OK. The electric charge simply describes the forces between electrically charged bodies. This would have a small time dependence — no problem.

Einstein: I could live with that too, assuming that the observations are correct. But going back to the 1.2×10^{-15} per year that you mentioned... your description implied that the change is still happening at the present time. But does it really have to be that way? Could it not be that the time variation stopped, say, three billion years ago?

Haller: Yes, that's true. But why should the time variation of α have stopped three billion years ago? Why not ten billion years ago?

In any case, if it did stop at any point, the variation of α with time must be highly nonlinear. Since we have no theory about e and its time variation, it could be so. Nobody knows.

Einstein: Now another question about the variation of α with time. We have already spoken about the grand unification of all forces. In these theories, the constants that describe the strengths of the interactions are mutually dependent. Would a variation in α mean that the constants for the strong and weak interactions are also time dependent?

Haller: Yes, I shall address this question now. We would obtain a time dependent fine structure constant if the strength of the unified interactions depends on time.

Einstein: In this case I would expect that the constant of the strong interaction also varies in time. That would imply, I think, that atomic masses change over time.

Haller: That depends on the specific theory. I have, with a doctoral student, looked at a simple model of unification. Our calculations showed that the change in the scale of the strong interaction could be quite large, around 40 times as large as the change in α.

Newton: But that would mean that the atomic masses should vary 40 times as much as α. This could be observed, couldn't it?

Haller: Yes, we shall come to that. But let me remind you that we arrived at the number 40 if the time dependence came from the constant of the unified interactions. If we consider instead a variation in the energy of unification and leave the interaction constant invariant, we obtain instead of 40 a negative number: -31. So when α gets smaller, as observed, the strength of the strong interaction should increase.

Newton: But in any case it should be possible to measure the effect in quantum optics experiments.

Haller: I wondered about this a while ago. I proposed an experiment that has since been done in Munich, and the results are interesting. But we've been in this pool long enough now. Let us take a break from our discussion. We can continue our conversation in the restaurant.

So the three physicists left the pool. Haller and Newton went to their room; meanwhile, Einstein took another walk around the institute and along the Pacific coastline. Then they met in the institute's cafeteria. They ordered steaks and a bottle of Cabernet from the Napa valley. After the meal, Einstein started the discussion again.

Einstein: We should go back to physics. You said that you had thought of an experiment probing the time dependence of the strong interaction.

Haller: Until recently, I didn't know very much about laser physics, and I must admit that I didn't think very much of it either. The new records in precision achieved by our quantum optics physicists in Bern and their colleagues in Munich did not impress me much. It was just atomic physics and quantum electrodynamics. No matter how precise their measurements became, to me it was old hat. But when I was thinking about the time dependency of the strong interaction, it became clear to me that with quantum optics one could discover something very interesting.

My colleagues in Bern referred me to Theodor Hänsch, a very good laser physicist at the University of Munich. He had moved to Munich some time ago, leaving his chair at Stanford University. I was planning a trip to Munich anyway and I met Hänsch two weeks later.

Hänsch explained to me that he had completed an experiment some time ago in which he had compared the rate of a caesium "clock" with the rate of hydrogen transitions. The caesium transitions depend on the magnetic moment of the atom's nucleus — they are hyperfine transitions. The transitions in hydrogen are simple atomic transitions which are independent of the nucleus. I got interested.

Einstein: I can imagine why you did. In a hyperfine transition, the magnetic moment of the nucleus comes into play. If anything happens to the strong interaction, the magnetic moment will change. The time measured by caesium atoms would then be different to that measured by hydrogen atoms — the two clocks would not be synchronized.

Haller: Yes, if the scale of the strong interaction were to change, the magnetic moment of the caesium nucleus would change. Hence the strength of the hyperfine transition would change. But nothing would happen in the case of the hydrogen transition. This leads us to expect that a disparity should develop between the rate of the hydrogen clock and the rate of the caesium clock. If the two are compared today and then once again tomorrow, we should see a small difference.

I explained this to Hänsch. He went to the blackboard and did a calculation. Quickly, he reached the conclusion that he might be able to see an effect.

Newton: How big is the effect you would expect to see?

Haller: The effect observed by the astrophysicists amounts to a fractional change in α of 1.2×10^{-15} per year. Multiplying this by 40, one gets about 5×10^{-14} per year for the strong interaction. That would cause a difference between the rate of the caesium clock and the rate of the hydrogen clock.

Hänsch looked at his experimental results and concluded that he could not say anything about effects smaller than 10^{-13} per year. But he also told me that he could easily repeat the experiment with greater precision. There was just one problem. Hänsch had not used a normal caesium clock in his experiment but the best caesium clock in the world, which is stationed in Paris and has the name "Pharao". It is a very precise clock and Hänsch would need it for his new experiment. He called the director of the institute in Paris and ordered the Pharao clock again.

The experiment was to start in four weeks. Hänsch had the advantage that his hydrogen transition was still operating, and so he had only to connect the Pharao clock. The clock came to Munich three weeks later. I traveled between Bern and Munich often in the following weeks. Sometimes I did shifts on the experiment.

Einstein: How long did it take you to find the effect?

Haller: Regretfully, we did not find anything. In the end, we were able to say that the change must be smaller than -0.9×10^{-15} per year, with an error margin of about 5.2×10^{-15}. We were expecting an effect of the order of 5×10^{-14}, which is a factor of ten larger than this limit, so the simplest model was probably excluded.

Einstein: The error that you mentioned is pretty big. Could the predicted effect be there after all?

Haller: It's not impossible, but I think that the chances are quite low.

Einstein: What about the experiment now? Is it still going?

Haller: No, the French needed the Pharao clock back. But Hänsch is preparing a new experiment that will give a smaller error. I should also mention that a similar experiment is being conducted at the National Institute of Standards and Technology in Boulder, Colorado.

Newton: You just told us that Hänsch found nothing at the level you expected. What does that mean for your theory?

Haller: The prediction of the theory is not unique. I said that a dependence of the interactions of the Standard Model on time could come about either from a time dependence of the energy of unification or from a time dependence of the constant of the unified interaction. The signs are reversed for the two cases.

Einstein: Couldn't it be that both effects happen? The two effects could partially cancel each other, thus the effect for the strong interaction could be smaller than you calculated for the simplest case.

Haller: Yes, that is quite possible. There are indeed theories that include a time dependency in both effects. If there were a partial cancellation, we could expect the effect that Hänsch is looking for to be around 5×10^{-15}. This is not ruled out by the experimental results so far, and that's why Hänsch is planning a better experiment.

Newton: That's a good idea. There is a real chance to find an effect.

Haller: I also fully support the search. I should also mention that a group in the Netherlands investigated the hydrogen transitions in very distant quasars with a telescope in Chile. They observed a small time variation of the ratio of the proton mass and the electron mass. If we keep the electron mass constant, this would imply a variation in the proton mass of 5×10^{-15}. Hänsch could find such an effect easily in the new experiment.

In Munich they plan to look for effects of the order of 10^{-17} per year. For example, hyperfine transitions in indium and silver will be examined.

With this sensitivity we could detect a time dependency in the scale parameters of the strong interaction of the order of 2×10^{-16} per year.

Einstein: Well, we will see. They might not find anything.

Haller: Sure, but if they do, it would be a very important discovery. Important discoveries are often made unexpectedly. But it's getting late. I suggest we go to our rooms. Tomorrow we are going to SLAC.

At the Stanford Linear Accelerator Center

The next morning, Haller had to go looking for Einstein. He found him sitting in the bath at the hot pool, engaged in conversation with a pretty woman. It took some time for Haller to drag Einstein away from his new acquaintance. Einstein had a weak spot for beautiful women. A grumbling Einstein was led away as Haller prepared to continue their drive towards Stanford.

Soon they were in the car, heading north. Haller knew of a good restaurant in the woods at Big Sur, and it was not long before they reached the place. They had an abundant breakfast. Newton ordered bacon with three eggs, toast, and hash browns — he liked the American breakfast.

After eating, they continued north. Soon they reached the Monterey peninsula. Haller made a small detour. They drove through a beautiful area; to their right was thick forest, to their left, the beach. Soon after leaving Monterey, they reached Santa Cruz. They were through the city in minutes, and they continued their journey northbound on Highway 1. Half an hour later, Haller left the road near the coast to drive down to a park by the ocean. They left the car and walked down to the beach, which was very broad at this point. There were high sand dunes all around.

Einstein liked the area. He wanted to stay for a while and take an extended walk. Haller and Newton agreed. First they had to cross a river. Newton couldn't swim and refused to wade through the water. In the end, Einstein and Haller carried him across.

Their stroll turned into a long walk along the wild beach. They did not get back to the car until about 1 PM. Haller then took them a couple of miles further north to a small place he remembered called Half Moon Bay. Here they were served a first-class lunch of Californian crabs with toast.

It was after two in the afternoon before they were back in the car. Now they were heading east on Highway 92, towards Stanford. They took the Skyline Boulevard through the coastal hills for a while, then went down a steep mountain road. They passed the small town of Woodside, nothing more than a collection of houses in the woods. After that, they arrived in Palo Alto. It was already late afternoon, and Haller decided that they shouldn't go to the Stanford Linear Accelerator Center (SLAC) until the next day. They rented three rooms in a motel on the El Camino Real, the main street of Palo Alto.

Since it was not late, Haller suggested driving over to San Francisco for dinner. Haller knew the restaurants of San Francisco very well. He headed straight for the port and they were soon sitting at a restaurant in Fisherman's Wharf enjoying roasted salmon and crab salad, washed down with a good white wine from the Napa valley.

After the meal, Haller took them across the Golden Gate Bridge. They stopped at a café in Sausalito, just after the bridge's end. From there they had a wonderful view of the city and of Alcatraz. The island of Alcatraz, just off the coast, had been the site of the famous prison. It was late at night by the time that they arrived back at the motel.

The following morning, Haller drove Einstein and Newton to University Avenue in the center of Palo Alto, where there is a coffee house that serves a very good breakfast. As they left, they passed through the wonderful campus of Stanford University before reaching Sand Hill Road, the road which leads to SLAC.

Haller drove past the entrance to SLAC and onto the highway nearby, heading south. He stopped on a bridge, even though that is not allowed, and told his colleagues to take a look down. Einstein and Newton gazed at the long linear building in astonishment. This was SLAC, the big linear accelerator.

Einstein: My God! What sort of monster is that?

Haller: That monster is the linear particle accelerator SLAC. It accelerates electrons or positrons until they are traveling at nearly the speed of light. The particles come from up there, and they fly into the big building down there, where they collide with a target such as a metal block.

Newton: You told us before that it was here at SLAC that quarks were observed inside nuclei for the first time. Can you remind me how that worked?

Haller: Yes, it happened in 1967 when I was still a student. The experimentalists here were firing electrons at nuclei. To their great surprise, they found that some of the electrons were scattered at very large angles. This had not been expected, especially given the results of some experiments carried out at Stanford earlier. In these experiments, electrons had been collided with protons. The probability of finding a high energy proton after the scattering had turned out to be rather low, and the same had been expected to be true for high energy collisions of electrons with nuclei. It turned out differently.

Einstein: Interesting — the electrons must have collided head on with the quarks inside the nuclei to be scattered at large angles. This is very similar to Rutherford's experiment in which the atomic nucleus was discovered. So the quarks were found in an analogous manner.

Haller: Indeed, that is exactly how the quarks were found. More precise experiments have even revealed the charges of the quarks. The results were as expected: $+\frac{2}{3}$ and $-\frac{1}{3}$, measured in units of the electric charge of the electron.

Before these experiments at SLAC were done, only a few particle physicists took quarks seriously. But that changed quickly with these results. At the time, I was myself at SLAC, but I drove down to Pasadena every second week and collaborated with Gell-Mann at Caltech.

To me it was clear that the experiments should be understood in the framework of the quark model, but to my surprise Gell-Mann did not share my opinion. He was skeptical, and I had to convince him.

But let's get back into the car before we have problems with highway patrol.

Haller went to the next exit, turned the car around, and drove back to Sand Hill Road. Soon they were at the entrance to SLAC. After a short phone call to the director, Haller drove directly to the parking lot in front of the theory building. Haller insisted that Einstein and Newton wear caps. If they didn't, they would surely be recognized immediately, particularly Einstein with his crown of white hair. Since the dress code in California is pretty relaxed, no one paid much attention to the two older gentlemen in baseball caps.

Haller took Einstein and Newton upstairs to the secretary. She had worked there since the '60s, and Haller knew her well. The secretary immediately showed them an office that happened to be free; this would serve as their temporary office. Haller called an experimentalist, Richard Taylor, whom he knew well and asked him when they could go to look at the experiment. It was possible right away, so Haller, Einstein, and Newton went to Taylor's office.

On the way, they passed a bulletin board that had, amongst other news, something about the Newton Institute in Cambridge, England. Newton's picture was on the poster. Haller pointed it out to his guests and commented that Newton looked better in reality than in the picture. Lucky that Newton was wearing a cap. Otherwise he would surely have been spotted.

When they reached Taylor's office, Haller introduced Einstein and Newton as friends from Los Angeles. Taylor jokingly said to Einstein that he looked like Einstein; Einstein stayed cool.

Taylor was one of the group of physicists that found the quarks inside nuclei in the 1960s. In 1990 he had obtained the Nobel Prize for the discovery. Taylor brought them to the experimental area. There was hardly anything left of the old SLAC experiments. Taylor explained to them where his detector had been located and how it had worked. Then they moved to the area where PEP, a new machine at SLAC, was being used in experiments. With this accelerator one produces new heavy particles that are made up of b quarks. The b quark is a heavy brother of the d quark.

After the tour, they drove with Taylor to a restaurant in nearby Woodside for lunch. Afterwards they returned to SLAC. In the afternoon, they continued their discussion about the fundamental constants.

Haller: I'm aware that we are going to be separated tomorrow, since both of you have a flight to catch. I am going to drive you to the airport in San Francisco. As we are nearing the end of our discussions, I would like to raise some ideas relating to the Big Bang that seem strange to me and probably will to you as well.

Einstein: As far as the natural constants are concerned, there is much that's strange already. Start, I'm curious.

Haller: Yes, indeed, there are a lot of things about the natural constants that remain a mystery, at least that is how it seems today.

But let me start by returning to one simple fact. Recall that for the (c, s) and (t, b) quark pairs, the quark with a charge of $+\frac{2}{3}$ is substantially heavier than the one with the $-\frac{1}{3}$ charge. So one would expect this to be true for the (u, d) system as well, but the u quark with charge $+\frac{2}{3}$ is lighter than the d quark with charge $-\frac{1}{3}$.

Einstein: Yes, and we noted that if it were the other way around, the proton would be heavier than the neutron. As a result, the proton would not be stable, and so we would not exist.

Haller: That's right. The lightest atom in the universe would be helium. Life as we know it would not be possible, not even for the simplest bacteria. The point, you see, is that for life as we know it to exist, the natural constants must have very specific values.

Something similar is true for the fine structure constant. Let us imagine a universe in which α is $\frac{1}{133}$ or $\frac{1}{140}$. One would think that this would be OK, but it's not. A slight change in the fine structure constant doesn't make a whole lot of difference to atoms, but it does have a big impact on complex biomolecules. Many of these complex molecules, the ones we need for life, would no longer exist.

Newton: Well, in this world there would be no Einstein or Newton, but maybe there would be some unexpected form of life, an Einstein worm, perhaps.

Haller: But I have my doubts about whether such a worm would have found the theory of relativity. What we want is life in the form known to us. It looks as if the natural constants necessary for stable matter — α, the scale of the strong interaction, the masses of the u and d quarks, and the electron mass — have been chosen precisely so that life is possible, and this is very strange.

Einstein: I would like to point out a possibility. The Big Bang happened 14 billion years ago. It might be that some of the constants, the quark masses, for example, were produced in the Big Bang. When the temperature of the universe was still very high, the masses of the light quarks played no role whatsoever. They only became important once the universe had cooled down. Maybe the quark masses were in constant flux and just took some random values when the universe cooled. Then, if the Big Bang were to be repeated, everything would turn out differently. The u mass could end up ten times as large as the d mass, for instance.

Haller: It could definitely have happened that way. The fact that we exist in the universe, if this is the case, is then a big accident. Nobody knows. But seeing my watch, I think it's time to start worrying about food. I suggest we go now to a restaurant in Palo Alto, I know a nice Chinese restaurant on the El Camino.

A Strange Big Bang

14

Within a few minutes they were on Sand Hill Road heading towards Palo Alto. They reached the Chinese restaurant after half an hour, and they continued their discussion there.

Newton: I would like to know more about the Big Bang. Why do physicists always talk about the Big Bang? Is it possible that there was more than one Big Bang, or infinitely many, or none at all?

Haller: Mr. Newton, you could be on the right track. We don't really know what happened in the Big Bang. Some physicists believe that time and space were created in the Big Bang, but they don't really know for sure how this might work.

But back to what is observed. More than a decade after you, Mr. Einstein, proposed your theory of gravity, astronomers observed that galaxies are drifting apart and that the velocity increases the further apart the objects are. Modern astronomical telescopes, such as the Hubble

Space Telescope, in orbit since 1990, have verified that this expansion includes the whole of the observable universe.

The Hubble Space Telescope is named after Edwin Hubble from Caltech, the man who discovered the cosmic expansion. You, Mr. Einstein, had the pleasure of meeting him often. With his telescope at Mt Wilson near Pasadena it was possible to see galaxies at distances of up to 5 billion light years away from us.

In the early phase of the universe, around 12 to 13 billion years ago, all the matter in the universe was concentrated in a small space. We believe today that the expansion of the cosmos is the result of an explosion, the Big Bang. The temperature must have been very high at the beginning. As a consequence of the afterglow of this cosmic fireball, the space between galaxies should still be filled with electromagnetic radiation. The idea of such radiation was first discussed by George Gamow after the Second World War. He even predicted the temperature of the radiation, calculating that it would be of the order of 10 Kelvin — very small, but not zero.

Nobody took Gamow's prediction seriously. The long-wavelength microwave radiation in intergalactic space was first discovered by physicists at the Bell research laboratories in Murray Hill; they stumbled upon it by accident in 1964. That this really was radiation that filled the whole cosmos was verified in the early 1990s by a satellite that had been put in orbit expressly to study this radiation. The satellite, called the Cosmic Microwave Background Explorer (COBE), was capable of measuring the temperature of the radiation with high precision. It is only 2.73° above absolute zero, not too far away from the temperature of 10° predicted by Gamow.

The microwave radiation, which, like light, is made up of photons, is a messenger from a time about 100,000 years after the Big Bang. At that time the universe had a temperature of 3000° and consisted of a plasma of nuclei and electrons. Through its expansion, the universe has cooled down to just a few Kelvin. COBE, as well as measuring the temperature

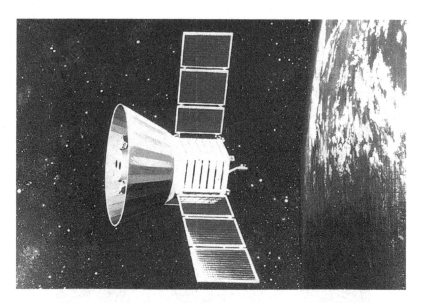

The COBE satellite.

of the cosmic radiation, was able to detect very small variations in the temperature across the sky, and these results allow us to draw conclusions about the distribution of matter in the very early cosmos.

If we wish to describe the behavior of the universe immediately after the Big Bang, it is necessary to understand the behavior of matter at extremely high temperatures and high densities. This leads us directly to nuclear physics and the physics of elementary particles. When particle physicists use accelerators to collide particles such as protons or electrons, they manufacture an extremely high energy density in a very tiny space, just as in the Big Bang.

Newton: Yes, but accelerators only create these conditions in a very small volume, not in the whole universe.

Haller: Yes, that is why the results of the accelerator experiments should be applied with caution. But, even so, insights from particle physics have enabled us to calculate details of the cosmic development after the Big Bang. It seems that immediately after the Big Bang the universe

The Big Bang.

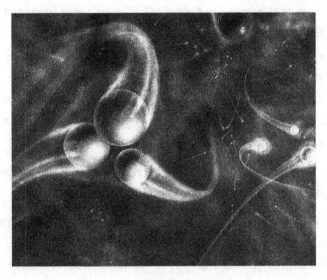

Three quarks form a nucleon.

consisted of a hot plasma of elementary particles. This plasma expanded and cooled down very quickly. Fractions of a second after the explosion, the first structures were formed.

Quarks came together in groups of three to form protons and neutrons. About one minute after the Big Bang helium nuclei formed through the combination of two neutrons and two protons. Calculations suggest that a short time after the Big Bang, about 24% of all the matter in the universe was in the form of helium nuclei. Measurements of the density of helium in the universe have confirmed this. Heavy elements, such as carbon and iron, formed later.

A few tens of thousand of years after the Big Bang, the first atoms were formed. Protons combined with electrons to produce hydrogen atoms. Thus the electrically neutral matter separated from the radiation, which interacts primarily with electrically charged particles. As the expansion continued, the first large clumps of matter condensed as a result of gravity — these clumps were the precursors of galaxies.

Since photons, like all other particles, are influenced by gravity, the formation of these clumps should lead to small temperature variations in

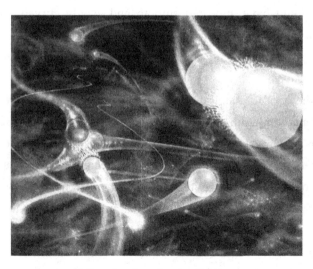

Helium forming after the Big Bang.

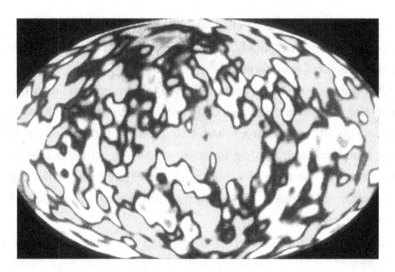

The fluctuation in cosmic temperature as measured by COBE.

the cosmic microwave background radiation. In 1992 the COBE satellite discovered such fluctuations. Judging from the size of the fluctuations, the original density variations were, at most, at the level of a mere hundred-thousandth of the density, so physicists concluded that the universe must have been very homogeneous and isotropic to begin with. At a quick glance it would have appeared smooth and evenly curved, just as the ocean appears smooth when seen from the window of an airplane on a trans-Atlantic flight. Only a closer look would reveal the unevenness caused by waves.

With the COBE satellite physicists discovered structures with a diameter one-hundredth the diameter of the observed universe. That is larger than the largest accumulations of matter visible today, the so-called superclusters of galaxies. Since then it has become possible, using very sensitive equipment, to detect temperature fluctuations in the background radiation that correspond to structures that are "only" about the size of a galactic cluster. New satellite experiments are being prepared. They aim to provide further knowledge on the accumulations of matter that led to the galaxies that we see today.

Many details of the structure formation are still unclear. The problem is closely connected to the question of whether there are other forms of matter besides the ones we know about. As far back as the 1930s scientists had noticed that gravitation in large galactic clusters was considerably stronger than expected from the amount of matter observed. Today we are certain that, in addition to normal matter, another form of matter exists. This matter constitutes up to 90% of all matter, but we can only see it through its gravitational effects — it is dark matter. What this dark matter is made of, we do not know.

Einstein: So a large fraction of the universe is made of some ghost material, and nobody knows what it is? That's not exactly satisfying. Actually, that's pretty frustrating!

Haller: Yes, it is frustrating. Perhaps neutrinos, the neutral partners of the electrons, might play a role. Hitherto undiscovered particles might also be involved. A candidate for the dark matter are heavy particles that several theoretical models predict, but which cannot yet be produced by accelerators. These particles would have to have formed in the hot soup right after the Big Bang and lead a ghost-like existence in today's universe. Particle physicists plan to look for such particles with the LHC, the new collider at CERN.

Newton: What was the cause of the Big Bang? Where did the tremendous energy out of which our universe developed come from?

Haller: Nobody knows. Possibly modern cosmology will eventually find answers to these questions. Such questions are closely intertwined with other puzzling aspects of the cosmos. The universe has been continually expanding since its creation around 14 billion years ago. The ratio of the observed matter density, including the dark matter, to the critical density at which the expansion would stop as a result of gravitation is in the vicinity of one. This ratio, generally known as omega (Ω), is an important number. The fact that Ω doesn't have a value like 100 or 0.01

is amazing. Whoever set the cosmic dynamics in motion must have paid pretty close attention to making the kinetic energy of the cosmic mass almost the same as its gravitational energy.

Another unsolved question is why the matter and the cosmic background radiation are homogeneously distributed in the observable universe. This is hard to understand. The turbulence of the Big Bang should still be seen today. A homogenization of the shockwaves associated with the Big Bang is not possible for the simple reason that these effects could not have spread faster than the speed of light. This would not be fast enough to homogenize the early universe, as one can see by considering the case of a galaxy that is five billion light years away from the Milky Way today. When the universe was one million years old, it was only a thousandth of its present size, therefore the distance between the two galaxies then was about five million light years. That would mean that the time that had passed since the Big Bang, one million years, would be too short for the two galaxies to have exchanged light signals. It follows that there could not have been any contact between the matter of one galaxy and that of the other. Why the universe is homogeneous remains incomprehensible.

Einstein: Once I introduced a cosmological constant. I then dropped it, but maybe this constant does indeed exist. It could very well have something to do with this problem.

Haller: Indeed, shortly after you proposed your theory of gravitation, you postulated a natural constant, the cosmological constant. This can be understood as a repulsive force that compensates the attractive force of gravity at great distances. If gravity could be switched off, the cosmological constant would result in a rapid inflation of the cosmos.

After it was discovered that the universe was expanding, the cosmological constant was not needed any more, and you dropped it.

Now particle physicists are considering again a cosmological constant. You will recall that a hypothetical field called the Higgs field was introduced into theories describing the dynamics of elementary particles. Its sole purpose is to give particles their mass. Simultaneously it breaks the symmetries of the particle interactions. This creation of mass can be pictured as a phase transition. It's like the freezing of water: the fluid water is homogeneous, but the rugged crystals of ice are not, so the symmetry of translation is broken.

The Higgs field, which we discussed, has the strange property that it can produce a cosmological constant. The field turns "empty" space into a space that has energy. However, the cosmological constant that particle physicists calculate from this model is far too large to play a role in the description of cosmic evolution.

At the beginning of the 1980s, physicists investigated the influence of the hypothetical Higgs field on cosmology. They came to the conclusion that the Higgs field played a decisive role in the Big Bang. According to this picture, the cosmological constant was very large right after the Big Bang. This would have led to a violent and extremely rapid expansion of the universe, known as inflation.

During the inflation a volume smaller than the nucleus of an atom expands rapidly until it is the size of a tennis ball. Then a cosmic phase transition sets in, freeing enormous amounts of energy. This is similar to the way that energy is set free when water freezes. This energy flash is the actual birth of our cosmos, the Big Bang.

After the phase transition, a normal expansion sets in. This is a slow and quiet process, unlike the period of inflation. The thing to note is that the universe was rather small before inflation, so regions that are far apart today could once have been in contact. The inflation would therefore have homogenized the universe, getting rid of all the cosmic crinkles. After the inflation the observable universe should be very homogeneous, which is indeed what we observe. The models of inflation predict that the ratio of the real to the critical mass density, the parameter Ω, mentioned before, should be equal to one.

Einstein: This sounds pretty good. My constant seems to have achieved something, at least it has made Ω equal to one.

The idea of a Big Bang as the result of a phase transition after inflation enables us to understand the observable cosmos as part of a much bigger system. It becomes possible to think of the Big Bang as a process that took place in a certain area of the cosmos, while other areas remained untouched. This would give real meaning to questions that have been dismissed in the past as nonsense — for instance, the question "what was before the Big Bang?" becomes meaningful. There could be a Big Bang in another region of the cosmos today. Maybe the cosmos is like fireworks, with creation and destruction processes constantly occurring, only we don't notice this because, thanks to inflation, we live in a relatively calm part of the universe.

Haller: Right, and then our world is just one of the many worlds that exist. In other regions of the cosmic space, other Big Bangs lead to other worlds. But it is not possible to go from one of these universes to another. The expression "universe" is not fitting any more; we speak of a multiverse.

Einstein: Now I see what you're getting at. In a multiverse Big Bangs take place constantly. The natural constants are formed accidentally in each Big Bang, so it is not surprising that we are living in a strange universe with weird values for the quark masses, a funny fine structure constant etc., a universe that just happens to have constants well suited for life. The natural constants were formed by chance. Of the many universes that are born, one will have the values we need for life, and that is the universe we live in.

Newton: If this is the case, it makes no sense to think about why, for example, the u quark mass is smaller than the d quark mass. That is just how it is in our world, by chance. In another world it might be different, but there would be nobody there to worry about it. We will never have a chance to understand why the u mass is smaller than the d mass, or why

The multiverse.

the fine structure constant is so close to the inverse of 137. The dream is over: everything is but a coincidence!

Haller: Yes, life in the way we know would probably not exist in most other universes. Some of them are likely to be pretty bare, they might not even have atoms. I wouldn't exclude the possibility that out of the many possible universes, probably infinitely many, our universe is the only one with some reasonable form of life.

Einstein: It could be that you are right, but I don't like it. I want a world where, in the end, everything can be calculated. But that's just my dream, and I might be wrong.

Haller: So, you see, this evening we have reached the frontier of current research. We should end our discussion here. Let us go back to the motel.

EPILOGUE

The next morning they had breakfast together at a nearby coffee shop. Then Haller drove Einstein and Newton to the San Francisco airport. Einstein and Newton were booked on a flight to Washington at 11 AM. Haller parked in the underground parking lot and brought his colleagues to the check-in counter.

Einstein: So, take care, Mr. Haller. We have had a very good discussion. The problem of the natural constants is a fundamental problem, but hopefully it can be solved some day. You and your colleagues have some tough work to do. When we meet again, you will have to tell me exactly what progress has been made. Maybe the LHC at CERN will tell us something new about the natural constants. In any case, goodbye, until we meet again.

Newton: I have enjoyed our discussion very much. These have been truly interesting days. Physics has become exciting. Until the next time, goodbye and all the best. Good luck, too, for the rest of your stay at Caltech. Perhaps you will find out there what dark matter is made of.

Haller: I will try my best, but I don't think that progress on dark matter will be made so quickly. I wish you a good flight. Don't forget to visit the Smithsonian while you are in Washington. It is a wonderful museum. I love it, and I visit it every time I am in Washington. Goodbye, I hope that we will meet again soon.

Haller walked slowly back to the parking lot. He was soon on the highway heading towards Palo Alto. He drove to SLAC where he was scheduled to give a seminar that afternoon.

❉ ❉ ❉

Haller felt the hand of the stewardess on his shoulder. He awoke and realized that he had been asleep for some time. The plane was already above the mountains of San Gabriel, to the north of Pasadena. Haller grinned at the thought of the dream he had just had: he had been discussing physics with Einstein and Newton, what a wonderful constellation.

He looked down at greater Los Angeles. Under him and to the left was the center of the city with its skyscrapers; Chinatown, with its restaurants, was right next door. He had been there often. To the right, he saw Hollywood Boulevard. Then the plane was above the campus of UCLA, another place he knew very well. In a couple of weeks he was to give a seminar there. The plane landed a few minutes later at the airport.

Haller rented a car and was soon on the highway heading towards Pasadena. An hour later he was in his room in the Athenaeum at Caltech. The lady receptionist, whom he had known for many years, gave him the big room on the first floor again. Einstein had once stayed in this room.

Before he went to dinner, Haller went to the small library of the Athenaeum. Nothing had changed; everything was just as it had been when Einstein gave a lecture here in 1929. Haller sat down and thought about the dream he had had on the plane.

Having Einstein and Newton here with him, like in the dream, would really be something. He thought, sadly, about Einstein's time at Caltech. Shortly after Einstein had been in Caltech, Hitler had taken over in Germany and Einstein could not return. He never saw his beloved summerhouse in Caputh again.

Haller went to the restaurant of the Athenaeum. He ordered a filet mignon and a bottle of Zinfandel from the Napa Valley. And he thought of Einstein and Newton and of the great discussions that they had enjoyed here in his dream.

INDEX